T0255265

Beginning Apache Pig

Big Data Processing Made Easy

Balaswamy Vaddeman

Apress®

Beginning Apache Pig: Big Data Processing Made Easy

Balaswamy Vaddeman
Hyderabad, Andhra Pradesh, India

ISBN-13 (pbk): 978-1-4842-2336-9   ISBN-13 (electronic): 978-1-4842-2337-6

DOI 10.1007/978-1-4842-2337-6

Library of Congress Control Number: 2016961514

Managing Director: Welmoed Spahr
Lead Editor: Celestin Suresh John
Technical Reviewer: Manoj R. Patil
Editorial Board: Steve Anglin, Pramila Balan, Laura Berendson, Aaron Black,
    Louise Corrigan, Jonathan Gennick, Robert Hutchinson, Celestin Suresh John,
    Nikhil Karkal, James Markham, Susan McDermott, Matthew Moodie, Natalie Pao,
    Gwenan Spearing
Coordinating Editor: Prachi Mehta
Copy Editor: Kim Wimpsett
Compositor: SPi Global
Indexer: SPi Global
Artist: SPi Global

Distributed to the book trade worldwide by Springer Science+Business Media New York, 233 Spring Street, 6th Floor, New York, NY 10013. Phone 1-800-SPRINGER, fax (201) 348-4505, e-mail orders-ny@springer-sbm.com, or visit www.springeronline.com. Apress Media, LLC is a California LLC and the sole member (owner) is Springer Science + Business Media Finance Inc (SSBM Finance Inc). SSBM Finance Inc is a **Delaware** corporation.

For information on translations, please e-mail rights@apress.com, or visit www.apress.com.

Apress and friends of ED books may be purchased in bulk for academic, corporate, or promotional use. eBook versions and licenses are also available for most titles. For more information, reference our Special Bulk Sales–eBook Licensing web page at www.apress.com/bulk-sales.

Any source code or other supplementary materials referenced by the author in this text are available to readers at www.apress.com. For detailed information about how to locate your book's source code, go to www.apress.com/source-code/. Readers can also access source code at SpringerLink in the Supplementary Material section for each chapter.

Printed on acid-free paper

The six most important people in my life:

The late Kammari Rangaswamy (Teacher)

The late Niranjanamma (Mother)

Devaiah (Father)

Radha (Wife)

Sai Nirupam (Son)

Nitya Maithreyi (Daughter)

Contents at a Glance

Contents

About the Author

Balaswamy Vaddeman is a thinker, blogger, and serious and self-motivated big data evangelist with 10 years of experience in IT and 5 years of experience in the big data space. His big data experience covers multiple areas such as analytical applications, product development, consulting, training, book reviews, hackathons, and mentoring. He has proven himself while delivering analytical applications in the retail, banking, and finance domains in three aspects (development, administration, and architecture) of Hadoop-related technologies. At a startup company, he developed a Hadoop-based product that was used for delivering analytical applications without writing code.

In 2013 Balaswamy won the Hadoop Hackathon event for Hyderabad conducted by Cloudwick Technologies. Being the top contributor at Stackoverflow.com, he helped countless people on big data topics at multiple web sites such as Stackoverflow.com and Quora.com. With so much passion on big data, he became an independent trainer and consultant so he could train hundreds of people and set up big data teams in several companies.

About the Technical Reviewer

Manoj R. Patil is a big data architect at TatvaSoft, an IT services and consulting firm. He has a bachelor's of engineering degree from COEP in Pune, India. He is a proven and highly skilled business intelligence professional with 17 years of information technology experience. He is a seasoned BI and big data consultant with exposure to all the leading platforms such as Java EE, .NET, LAMP, and so on. In addition to authoring a book on Pentaho and big data, he believes in knowledge sharing, keeps himself busy in corporate training, and is a passionate teacher. He can be reached at on Twitter @manojrpatil and at https://in.linkedin.com/in/manojrpatil on LinkedIn.

Manoj would like to thank his family, especially his two beautiful daughters, Ayushee and Ananyaa, for their patience during the review process.

Acknowledgments

Writing a book requires a great team. Fortunately, I had a great team for my first project. I am deeply indebted to them for making this project reality.

I would like to thank the publisher, Apress, for providing this opportunity.

Special thanks to Celestin Suresh John for building confidence in me in the initial stages of this project.

Special thanks to Subha Srikant for your valuable feedback. This project would have not been in this shape without you. In fact, I have learned many things from you that could be useful for my future projects also.

Thank you, Manoj R. Patil, for providing valuable technical feedback. Your contribution added a lot of value to this project.

Thank you, Dinesh Kumar, for your valuable time.

Last but not least, thank you, Prachi Mehta, for your prompt coordination.

CHAPTER 1

MapReduce and Its Abstractions

In this chapter, you will learn about the technologies that existed before Apache Hadoop, about how Hadoop has addressed the limitations of those technologies, and about the new developments since Hadoop was released.

Data consists of facts collected for analysis. Every business collects data to understand their business and to take action accordingly. In fact, businesses will fall behind their competition if they do not act upon data in a timely manner. Because the number of applications, devices, and users is increasing, data is growing exponentially. Terabytes and petabytes of data have become the norm. Therefore, you need better data management tools for this large amount of data.

Data can be classified into these three types:

- *Small data*: Data is considered small data if it can be measured in gigabytes.

- *Big data*: Big data is characterized by volume, velocity, and variety. *Volume* refers to the size of data, such as terabytes and more. *Velocity* refers to the age of data, such as real-time, near-real-time, and streaming data. *Variety* talks about types of data; there are mainly three types of data: structured, semistructured, and unstructured.

- *Fast data*: Fast data is a type of big data that is useful for the real-time presentation of data. Because of the huge demand for real-time or near-real-time data, fast data is evolving in a separate and unique space.

Small Data Processing

Many tools and technologies are available for processing small data. You can use languages such as Python, Perl, and Java, and you can use relational database management systems (RDBMSs) such as Oracle, MySQL, and Postgres. You can even use data warehousing tools and extract/transform/load (ETL) tools. In this section, I will discuss how small data processing is done.

Electronic supplementary material The online version of this chapter (doi:10.1007/978-1-4842-2337-6_1) contains supplementary material, which is available to authorized users.

© Balaswamy Vaddeman 2016
B. Vaddeman, *Beginning Apache Pig*, DOI 10.1007/978-1-4842-2337-6_1

Assume you have the following text in a file called `fruits`:

```
Apple, grape
Apple, grape, pear
Apple, orange
```

Let's write a program in a shell script that first filters out the word *pear* and then counts the number of words in the file. Here's the code:

```
cat fruits|tr ',' '\n'|grep -v -i 'pear'|sort -f|uniq  -c -i
```

This code is explained in the following paragraphs.

In this code, `tr` (for "translate" or "transliterate") is a Unix program that takes two inputs and replaces the first set of characters with the second set of characters. In the previous program, the `tr` program replaces each comma (,) with a new line character (\n). `grep` is a command used for searching for specific text. So, the previous program performs an inverse search on the word *pear* using the -v option and ignores the case using -i.

The `sort` command produces data in sorted order. The -f option ignores case while sorting.

`uniq` is a Unix program that combines adjacent lines from the input file for reporting purposes. In the previous program, `uniq` takes sorted words from the `sort` command output and generates the word count. The -c option is for the count, and the -i option is for ignoring case.

The program produces the following output:

```
Apple 3
Grape 2
Orange 1
```

You can divide program functionality into two stages; first is tokenize and filtering, and second is aggregation. Sort is supporting functionality of aggregation. Figure 1-1 shows the program flow.

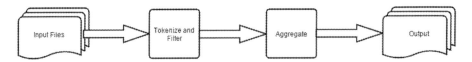

Figure 1-1. *Program flow*

The previous program can be run on a single machine and on small data. Such simple programs can be used to perform simple operations such as searching and sorting on one file at a time. However, writing complex queries involving multiple files and multiple conditions requires better data processing tools. Database management systems (DBMS) and RDBMS technologies were developed to address querying problems with structured data.

Relational Database Management Systems

RDBMSs were developed based on the relational model founded by E. F. Codd. There are many commercial RDBMS products such as Oracle, SQL Server, and DB2. Many open source RDBMSs such as MySQL, Postgres, and SQLite are also popular. RDBMSs store data in tables, and you can define relations between tables.

Here are some advantages of RDBMSs:

- RDBMS products come with sophisticated query languages that can easily retrieve data from multiple tables with multiple conditions.

- The query language used in RDBMSs is called Structured Query Language (SQL); it provides easy data definition, manipulation, and control.

- RDBMSs also support transactions.

- RDBMSs support low-latency queries so users can access databases interactively, and they are also useful for online transaction processing (OLTP).

RDBMSs have these disadvantages:

- As data is stored in table format, RDBMSs support only structured data.

- You need to define a schema at the time of loading data.

- RDBMSs can scale only to gigabytes of data, and they are mainly designed for frequent updates.

Because the data size in today's organizations has grown exponentially, RDBMSs have not been able to scale with respect to data size. Processing terabytes of data can take days.

Having terabytes of data has become the norm for almost all businesses. And new data types like semistructured and unstructured have arrived. Semistructured data has a partial structure like in web server log files, and it needs to be parsed like Extensible Markup Language (XML) in order to analyze it. Unstructured data does not have any structure; this includes images, videos, and e-books.

Data Warehouse Systems

Data warehouse systems were introduced to address the problems of RDBMSs. Data warehouse systems such as Teradata are able to scale up to terabytes of data, and they are mainly used for OLAP use cases.

Data warehousing systems have these disadvantages:

- Data warehouse systems are a costly solution.

- They still cannot process other data types such as semistructured and unstructured data.

- They cannot scale to petabytes and beyond.

All traditional data-processing technologies experience a couple of common problems: storage and performance.

Computing infrastructure can face the problem of node failures. Data needs to be available irrespective of node failures, and storage systems should be able to store large volumes of data.

Traditional data processing technologies used a scale-up approach to process a large volume of data. A scale-up approach adds more computing power to existing nodes, so it cannot scale to petabytes and more because the rest of computing infrastructure becomes a performance bottleneck.

Growing storage and processing needs have created a need for new technologies such as parallel computing technologies.

Parallel Computing

The following are several parallel computing technologies.

GFS and MapReduce

Google has created two parallel computing technologies to address the storage and processing problems of big data. They are Google File System (GFS) and MapReduce. GFS is a distributed file system that provides fault tolerance and high performance on commodity hardware. GFS follows a master-slave architecture. The master is called Master, and the slave is called ChunkServer in GFS. MapReduce is an algorithm based on key-value pairs used for processing a huge amount of data on commodity hardware. These are two successful parallel computing technologies that address the storage and processing limitations of big data.

Apache Hadoop

Apache Hadoop is an open source framework used for storing and processing large data sets on commodity hardware in a fault-tolerant manner.

Hadoop was written by Doug Cutting and Mark Cafarella in 2006 while working for Yahoo to improve the performance of the Nutch search engine. Cutting named it after his son's stuffed elephant toy. In 2007, it was given to the Apache Software Foundation.

Initially Hadoop was adopted by Yahoo and, later, by companies like Facebook and Microsoft. Yahoo has about 100,000 CPUs and 40,000 nodes for Hadoop. The largest Hadoop cluster has about 4,500 nodes. Yahoo runs about 850,000 Hadoop jobs every day. Unlike conventional parallel computing technologies, Hadoop follows a scale-out strategy, which makes it more scalable. In fact, Apache Hadoop had set a benchmark by sorting 1.42 terabytes per minute.

Most of Hadoop is written in Java, but it has support for many programming languages such as C, C++, Python, and Scala through its streaming module. Apache Hadoop was initially written for high throughput and batch-processing systems. RDBMS technologies were written for frequent modifications in data, whereas Hadoop has been written for frequent reads.

Moore's law says the processing capability of a machine will double every two years. Kryder's law says the storage capacity of disks will grow faster than Moore's law. The cost of computing and storage devices will go down every day, and these two factors can support more scalable technologies. Apache Hadoop was designed while keeping these things in mind, and parallel computing technologies like this will become more common going forward.

The latest Apache Hadoop contains three modules, as shown in Figure 1-2. They are HDFS, MapReduce, and Yet Another Resource Negotiator (YARN).

Figure 1-2. *The three components of Hadroop*

HDFS

The Hadoop distributed file system is used for storing large data sets. It divides files into blocks and stores every block on at least multiple nodes. This is called a *replication factor*, and by default it is 3. HDFS is fault-tolerant because it has more than one replica for every block, so it can handle node failures without affecting data processing. A block of HDFS is the same as an operating system block, but a HDFS block size is larger, such as 64 MB or 128 MB. Unlike traditional storage systems, it is highly scalable. It does not require any special hardware and can work on commodity hardware.

Assume you have a replication factor of 3, a block size of 64 MB, and 640 MB of data needs to be uploaded into HDFS. At the time of uploading the data into HDFS, 640 MB is divided into 10 blocks with respect to block size. Every block is stored on three nodes, which would consume 1920 MB of space on a cluster.

HDFS follows a master-slave architecture. The master is called the *name node*, and the slave is called a *data node*. The data node is fault tolerant because the same block is replicated to two more nodes. The name node was a single point of failure in initial versions; in fact, Hadoop used to go down if the name node crashed. But Hadoop 2.0+ versions have high availability of the name node. If the active name node is down, the standby name node becomes active without affecting the running jobs.

MapReduce

MapReduce is key-value programming model used for processing large data sets. It has two core functions: Map and Reduce. They are derived from functional programming languages. Both functions take a key-value pair as input and generate a key-value pair as output.

The Map task is responsible for filtering operations and preparing the data required for the Reduce tasks. The Map task will generate intermediate output and write it to the hard disk. For every key that is being generated by the Map task, a Reduce node is identified and will be sent to the key for further processing.

The Map task takes the key-value pair as input and generates the key-value pair as output.

```
(key1, value1) ----------------> Map Task----------------> (Key2, Valu2)
```

The Reduce task is responsible for data aggregation operations such as count, max, min, average, and so on. A reduce operation will be performed on a per-key basis. Every functionality can be expressed in MapReduce.

The Reduce task takes the key and list of values as input and generates the key and value as output.

```
(key2, List (value2))--------> Reduce Task ---------------> (Key3, value3)
```

In addition to the Map and Reduce tasks, there is an extra stage called the *combiner* to improve the performance of MapReduce. The combiner will do partial aggregation on the Map side so that the Map stage has to write less data to disk.

You will now see how MapReduce generates a word count. Figure 1-3 depicts how MapReduce generates the fruits word count after filtering out the word *pear*.

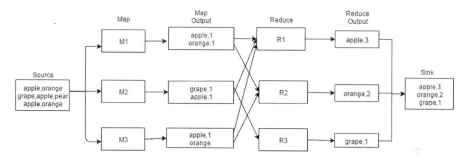

Figure 1-3. *MapReduce generating a word count*

Source and Sink are HDFS directories. When you upload data to HDFS, data is divided into chunks called *blocks*. Blocks will be processed in a parallel manner on all available nodes.

The first stage is Map, which performs filtering and data preparation after tokenization. All Map tasks (M1, M2, and M3) will do the initial numbering for words that are useful for the final aggregation. And M2 filters out the word *pear*.

The key and list of its values are retrieved from the Map output and sent to the reducer node. For example, the Apple key and its values (1, 1, 1) are sent to the reducer node R1. The reducer aggregates all values to generate the count output.

Between Map and Reduce, there is an internal stage called *shuffling* where the reducer node for the map output is identified.

You will now see how to write the same word count program using MapReduce. You first need to write a mapper class for the Map stage.

Writing a Map Class

The following is the Map program that is used for the same tokenization and data filtering as in the shell script discussed earlier:

```
import java.io.IOException;
import java.util.StringTokenizer;
import org.apache.hadoop.io.IntWritable;
import org.apache.hadoop.io.LongWritable;
import org.apache.hadoop.io.Text;
import org.apache.hadoop.mapreduce.Mapper;
public class FilterMapper extends Mapper<LongWritable, Text, Text, IntWritable> {
    private final static IntWritable one = new IntWritable(1);
    private Text word = new Text();
    public void map(LongWritable offset, Text line, Context context) throws
    IOException, InterruptedException {
        //tokenize line with comma as delimiter
        StringTokenizer itr = new StringTokenizer(line.toString(),",");
        //Iterate all tokens and filter pear word
        while (itr.hasMoreTokens()) {
```

```
            String strToken=itr.nextToken();
            if(!strToken.equals("pear")){
//converting string data type to text data type of mapreduce

            word.set(strToken);
            context.write(word, one);//Map output
            }


      }
    }
}
```

The Map class should extend the Mapper class, which has parameters for the input key, input value, output key, and output value. You need to override the map() method. This code specifies LongWritable for the input key, Text for the input value, Text for the output key, and IntWritable for the output value.

In the map() method, you use StringTokenizer to convert a sentence into words. You are iterating words using a while loop, and you are filtering the word *pear* using an if loop. The Map stage output is written to context.

For every run of the map() method, the line offset value is the input key, the line is the input value, the word in the line will become an output key, and 1 is the output value, as shown in Figure 1-4.

Figure 1-4. *M2 stage*

The map() method runs once per every line. It tokenizes the line into words, and it filters the word *pear* before writing other words with the default of 1.

If the combiner is available, the combiner is run before the Reduce stage. Every Map task will have a combiner task that will produce aggregated output. Assume you have two *apple* words in the second line that is processed by the M2 map task.

The Map output without the combiner will look like Figure 1-5.

Figure 1-5. *Map output without the combiner*

Even combiner follows the key-value paradigm. Like the Map and Reduce stages, it will have an input key and input value and also an output key and output value. The combiner will write its output data to disk after aggregating the map output data. The combiner will write relatively less data to disk as it is aggregated, and less data is shuffled to the Reduce stage. Both these things will improve the performance of MapReduce.

Figure 1-6 shows the combiner writing aggregated data that is apple,2 here.

Figure 1-6. *The combiner writing aggregated data*

Writing a Reduce Class

The following is a reducer program that does the word count on the map output and runs after the Map stage if the combiner is not available:

```
import java.io.IOException;
import org.apache.hadoop.io.IntWritable;
import org.apache.hadoop.io.Text;
import org.apache.hadoop.mapreduce.Reducer;

public class WordCountReducer extends Reducer<Text, IntWritable, Text,
IntWritable> {
    private IntWritable count = new IntWritable();

    public void reduce(Text word, Iterable<IntWritable> values, Context context)
            throws IOException, InterruptedException {
        int sum = 0;
        // add all values for a key i.e. word
        for (IntWritable val : values) {
            sum += val.get();
        }
        count.set(sum);//type cast from int to IntWritable
context.write(word, count);
    }
}
```

The reduce class should extend the Reducer class, which has parameters for the input key, input value, output key, and output value. You need to override the reduce() method, and you specify the text data type for the input key and IntWritable for the input value, and these two should match with the map output key and value data types. You also specify the output key as text and the output value as IntWritable.

For every run of reducer, the Map output key and its list of values are passed to the reduce() method as input. The list of values is iterated using a for loop because they are already iterable. Using the get() method of IntWritable, you get the value of the Java int data type that you would add to the sum variable. After completing the reduce() method for the partial word key, the word and count are generated as the reducer output. The reduce() method is run once per key, and the Reduce stage output is written to context just like map output. Figure 1-7 shows *apple* and the list of values (1,1,1) processed by Reduce task R2.

Figure 1-7. *Reduce task R2*

Writing a main Class

The following is the main class that generates the word count using the mapper class and the reducer class:

```
import org.apache.hadoop.conf.Configuration;
import org.apache.hadoop.fs.Path;
import org.apache.hadoop.io.IntWritable;
import org.apache.hadoop.io.Text;
import org.apache.hadoop.mapreduce.Job;
import org.apache.hadoop.mapreduce.lib.input.FileInputFormat;
import org.apache.hadoop.mapreduce.lib.output.FileOutputFormat;

public class WordCount {

    public static void main(String[] args) throws Exception {
        Configuration conf = new Configuration();
        Job job = Job.getInstance(conf, "word count");
        job.setJarByClass(WordCount.class);
        job.setMapperClass(FilterMapper.class);
        job.setReducerClass(WordCountReducer.class);
        job.setOutputKeyClass(Text.class);
        job.setOutputValueClass(IntWritable.class);
        //take input path from command line
        FileInputFormat.addInputPath(job, new Path(args[0]));
        //take output path from command line
        FileOutputFormat.setOutputPath(job, new Path(args[1]));
        System.exit(job.waitForCompletion(true) ? 0 : 1);
    }
}
```

In this code, you create a configuration object, and you pass it to the Job class.

You first pass the main class name in setJarByClass so that the framework will start executing from that class. You set the mapper class and the reducer class on the job object using the setMapperClass and setReducerClass methods.

FileInputFormat says the input format is available as a normal file. And you are passing the job object and input path to it. FileOutputFormat says the output format is available as a normal file. And you are passing the job object and output path to it. FileInputFormat and FileOutputFormat are generic classes and will handle any file type, including text, image, XML, and so on. You need to use different classes for handling different data formats. TextInputFormat and TextOutputFormat will handle only text data. If you want to handle binary format data, you need to use Sequencefileinputformat and sequencefileoutputformat.

If you want to specify key and value data types, you can control them from this program for both the mapper and the reducer.

Running a MapReduce Program

You need to create a .jar file with the previous three programs. You can generate a .jar file using the Eclipse export option. If you are creating a .jar file on other platforms like Windows, you need to transfer this .jar file to one of the nodes in the Hadoop cluster using FTP software such as FileZilla or WinScp. Once the JAR is available on the Hadoop cluster, you can use the Hadoop jar command to run the MapReduce program, like so:

```
Hadoop jar /path/to/wordcount.jar Mainclass InputDir OutputDir
```

Most grid computing technologies send data to code for processing. Hadoop works in the opposite way; it sends code to data. Once the previous command is submitted, the Java code is sent to all data nodes, and they will start processing data in a parallel manner. The final output is written to files in the output directory, and by default the job will fail if the output directory already exists. The total number of files will depend on the number of reducers.

1. Prepare data that is suitable for the combiner and write a program for word count using MapReduce that includes the combiner stage.

YARN

In earlier Hadoop versions, MapReduce was responsible for data processing, resource management, and scheduling. In Hadoop 2.0, resource management and scheduling have been separated from MapReduce to create a new service called YARN. With YARN, several applications such as in-memory computing systems and graph applications can co-exist with MapReduce.

YARN has a couple of important daemons. They are the resource manager and node manager. The resource manager is responsible for providing resources to all applications in the system. The node manager is the per-machine framework agent that is responsible for containers, monitoring their resource usage (CPU, memory, disk, network), and reporting the same to the resource manager. The per-application

11

application master is, in effect, a framework-specific library and is tasked with negotiating resources from the resource manager and working with the node manager to execute and monitor the tasks.

Benefits

Now that you know about the three components that make up Apache Hadoop, here are the benefits of Hadoop:

- Because Apache Hadoop is an open source software framework, it is a cost-effective solution for processing big data. It also runs on commodity hardware.

- Hadoop does not impose a schema on load; it requires a schema only while reading. So, it can process all types of data, that is, structured, semistructured, and unstructured data.

- Hadoop is scalable to thousands of machines and can process data the size of petabytes and beyond.

- Node failures are normal in a big cluster, and Hadoop is fault tolerant. It can reschedule failed computations.

- Apache Hadoop is smart parallel computing framework that sends code to data rather than data to code. This code-to-data approach consumes fewer network resources.

Use Cases

Initially Hadoop was developed as a batch-processing framework, but after YARN, Hadoop started supporting all types of applications such as in-memory and graph applications.

- Yahoo uses Hadoop in web searches and advertising areas.

- Twitter has been using Hadoop for log analysis and tweet analysis.

- Hadoop had been used widely for image processing and video analytics.

- Many financial companies are using Hadoop for churn analysis, fraud detection, and trend analytics.

- Many predictive analytics have been done using Hadoop in the healthcare industry.

- LinkedIn's "People You May Know" feature is implemented by using Apache Hadoop.

Problems with MapReduce

MapReduce is a low-level API. You need to think in terms of the key and value every time you use it. In addition, MapReduce has a lengthy development time. It cannot be used for ad hoc purposes. You need MapReduce abstractions, which hide the key-value programming paradigm from the user.

Chris Wensel addressed this problem by creating the Java-based MapReduce abstraction called Cascading.

Cascading

Cascading is a Java-based MapReduce abstraction used for building big data applications. It hides the key-value complexity of MapReduce from the programmer so that the programmer can focus on the business logic, unlike MapReduce. Cascading also has an API that provides several built-in analytics functions. You do not need to write functions such as count, max, and average, unlike MapReduce. It also provides an API for integration and scheduling apart from processing.

Cascading is based on a metaphor called pipes and filters. Basically, Cascading allows you to define a pipeline that contains a list of pipes. Once the pipe output is passed as an input to another pipeline, the pipelines will merge, join, group, and split the data apart from performing other operations on data. The pipeline will read data from the source tap and will write to the sink tap. The source tap, sink tap, and their pipeline are defined as a flow in Cascading. Figure 1-8 shows a sample flow of Cascading.

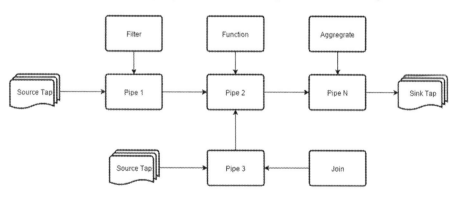

Figure 1-8. *Sample flow of Cascading*

Here is how to write a word count program in Cascading:

```
import java.util.Properties;
import cascading.flow.Flow;
import cascading.flow.local.LocalFlowConnector;
import cascading.operation.aggregator.Count;
import cascading.operation.regex.RegexSplitGenerator;
import cascading.pipe.Each;
```

```
import cascading.pipe.Every;
import cascading.pipe.GroupBy;
import cascading.pipe.Pipe;
import cascading.property.AppProps;
import cascading.scheme.local.TextLine;
import cascading.tap.SinkMode;
import cascading.tap.Tap;
import cascading.tap.local.FileTap;
import cascading.tuple.Fields;
public class WordCount {
 public static void main(String[] args) {
  Tap srcTap = new FileTap( new TextLine( new Fields(new String[]{"line"})) ,
               args[0] );
  Tap sinkTap =new FileTap(  new TextLine( new Fields(new String[]{"word" ,
               "count"})), args[1], SinkMode.REPLACE );
  Pipe words=new Each("start",new RegexSplitGenerator(","));
  Pipe group=new GroupBy(words);
  Count count=new Count();
  Pipe wcount=new Every(group, count);
  Properties properties = new Properties();
  AppProps.setApplicationJarClass( properties, WordCount.class );
  LocalFlowConnector flowConnector = new LocalFlowConnector();
  Flow flow = flowConnector.connect( "wordcount", srcTap, sinkTap, wcount );
  flow.complete();
 }
}
```

In this code, the Fields class is used for defining column names. TextLine will hold field names and also data path details. srcTap will have source field names and input path. snkTap will define output field names and the output path. FileTap is used to read data from the local file system. You can use HFS to run it on the HDFS data. SinkMode. REPLACE will replace the output data if it already exists in the output directory specified.

Each operator is allowed to perform an operation on each line. Here you are using the RegexSplitGenerator function that splits every line of text into words using a comma (,) as the delimiter. You are defining this pipe as words.

The GroupBy class works on the words *pipe* to arrange words into groups and creates a new pipe called group. Later you will create a new pipe account that will apply the count operation on every group using the Every operator.

Properties allow you to provide values to properties. You are not setting any properties. You will use the put() method to insert property values.

```
properties.put("mapred.reduce.tasks", -1);
```

You will create an object for LocalFlowConnector and will define the flow mentioning the source tap and sink tap and last pipe. The functionality of the application will be resolved starting from the last pipe to the first pipe.

`LocalFlowConnector` will help you to create a local flow that can be run on the local file system. You can use `HadoopFlowConnector` for creating a flow that works on the Hadoop file system. `flow.complete()` will start executing the flow.

1. Modify the previous Cascading program to filter the word *pear*.

Benefits

These are the benefits of Cascading:

- Like MapReduce, it can process all types of data, such as structured, semistructured, and unstructured data.

- Though it is a MapReduce abstraction, it is still easy to extend it.

- You can rapidly build big data applications using Cascading.

- Cascading is unit-testable.

- It follows fail-fast behavior, so it is easy to troubleshoot problems.

- It is proven as an enterprise tool and can seamlessly integrate with data-processing tools.

Use Cases

Cascading can be used as an ETL, batch-processing, machine-learning, and big data product development tool. Cascading can be used in many industries such as social media, healthcare, finance, and telecom.

Apache Hive

A traditional warehouse system is an expensive solution that will not scale to big data. Facebook has created a warehouse solution called Hive. Hive is built on top of Hadoop to simplify big data processing for business intelligence users and tools. The SQL interface in Hive has made it widely adopted both within Facebook and even outside of Facebook, especially after it was provided as open source to the Apache Software Foundation. It supports indexing and ACID properties.

Hive has some useful components such as the metastore, Hive Query Language, HCatalog, and Hive Server.

- The metastore stores table metadata and stats in an RDBMS such as MySQL, Postgres, or Oracle. By default it stores metadata in the embedded RDBMS Apache Derby.

- The Hive Query Language (HQL) is a SQL interface to Hadoop that is compiled into MapReduce code. Queries can be submitted through the command-line interface (CLI), the web interface, a Thrift client, an ODBC interface, or a JDBC interface. HQL can launch not only MapReduce but also Tez and Spark jobs.

- HCatalog is table and storage management tool that enables big data processing tools to easily read and write data.

- HiveServer2 is a Thrift client that enables BI tools to connect to Hive and retrieve results.

Here is how to write a word count program in Apache Hive:

```
select word,count(word) as count
from
(SELECT explode(split(sentence, ',')) AS word FROM texttable)temp
group by word
```

This writes a Hive query that filters the word *pear* and generates the word count.

split is used to tokenize sentences into words after applying a comma as a delimiter. explode is a table-generating function that converts every line of words into rows and names new column data as words. This creates a new temporary table called temp, generates a word-wise count using the group by and count functions from the temp table, and creates an alias called count. This query output is displayed on the console. You can create a new table from this table by prepending the create table as select statement like below.

```
Create table wordcount as
```

Benefits

Hive is a scalable data warehousing system. Building a Hive team is easy because of its SQL interface. Unlike MapReduce, it is suitable for ad hoc querying. With many BI tools available on top of Hive, people without much programming experience can get insights from big data. It can easily be extensible using user-defined functions (UDFs). You can easily optimize code and also support several data formats such as text, sequence, RC, and ORC.

Use Cases

Because Hive has a SQL interface, it was a quickly adopted Hadoop abstraction in businesses. Apache Hive is used in data mining, R&D, ETL, machine learning, and reporting areas. Many business intelligence tools provide facilities to connect to a Hive warehouse. Some tools include Teradata, Aster data, Tableau, and Cognos.

Apache Pig

Pig is a platform for analyzing large data sets with a sophisticated environment for optimization and debugging. It introduced a scripting-based language called Pig Latin that is used for data processing. Pig Latin is data flow language that follows a step-by-step process to analyze data. Pig Latin can launch MapReduce, Tez, and Spark jobs. Pig's current version is 0.15, and Pig support for Spark is a work in progress. Pig Latin can call Java, JavaScript, Python, Ruby, or Groovy code through UDFs.

It was developed by a team at Yahoo for researchers around 2006. In 2007, it was open sourced to the Apache Software Foundation. The purpose of Pig was to enable ad hoc querying on Apache Hadoop.

Here is how to write a word count program in Apache Pig:

```
input = LOAD '/path/to/input/file/' AS (line:Chararray);
Words = FOREACH input GENERATE FLATTEN(TOKENIZE(line,',')) AS word;
Grouped = GROUP words BY word;
wordcount = FOREACH Grouped GENERATE group, COUNT(word) as wordcount;
store wordcount into '/path/to/output/dir';
```

The load operator reads the data from the specified path after applying the schema that is specified after the As word. Here line is the column name, and chararray is the data type. You are creating a relation called input.

The FOREACH processes line by line on the relation input, and generate applies the Tokenize and Flatten functions to convert sentences into plain words using a comma delimiter, and the column name is specified as word. These words are stored in a relation called words. Words are arranged into groups using the Group operator. The next line is applied on a relation called grouped that performs the count function on every group of words. You are defining the column name as wordcount. You will store the final output in another directory using the store operator. The dump operator can be used for printing the output on the console.

1. Change the previous program to filter the word *pear*.

Pig Latin code can be submitted using its CLI and even using the HUE user interface. Oozie can use Pig Latin code as part of its workflow, and Falcon can use it as part of feed management.

Pig vs. Other Tools

The Hadoop ecosystem has many MapReduce abstractions, as shown in Figure 1-9. You will learn how Apache Pig is compared against others.

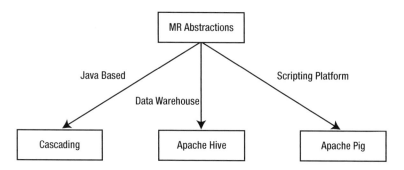

Figure 1-9. *Pig versus other tools*

MapReduce

MapReduce is a low-level API. Development efforts are required for even simple analytical functions. For example, joining data sets is difficult in MapReduce. Maintainability and reuse of code are also difficult in MapReduce. Because of a lengthy development time, it is not suitable for ad hoc querying. MapReduce requires a learning curve because of its key-value programming complexity. Optimization requires many lines of code in MapReduce.

Apache Pig is easy to use and simple to learn. It requires less development time and is suitable for ad hoc querying. A simple word count in MapReduce might take around 60 lines of code. But Pig Latin can do it within five lines of code. You can easily optimize code in Apache Pig. Unlike MapReduce, you just need to specify two relations and their keys for joining two data sets.

Cascading

Cascading extensions are available in different languages. Scalding is Scala-based, Cascalog is Clojure-based, and PyCascading is Python-based. All are programming language–based. Though it takes relatively less development time than MapReduce, it cannot be used for ad hoc querying. In Pig Latin, the programming language is required only for advanced analytics, not for simple functions. The word count program in Cascading requires 30 lines of code, and Pig requires only five lines of code. Cascading's pipeline will look similar to the data flow conceptually.

Apache Hive

Apache Hive was written only for a warehouse use case that can process only structured data that is available within tables. And it is declarative language that talks about what to achieve rather than how to achieve. Complex business applications require several lines of code that might have several subqueries inside. These queries are difficult to understand and are difficult to troubleshoot in case of issues. Query execution in Hive is from the innermost query to the outermost query. In addition, it takes time to understand the functionality of the query.

Pig is procedural language. Pig Latin can process all types of data: structured, semistructured, and unstructured including nested data. One of the main features of Pig is debugging. A developer can easily debug Pig Latin programs. Pig has a framework called Penny that is useful for monitoring and debugging Pig Latin jobs. The data flow language Pig Latin is written in step-by-step manner that is natural and easy to understand.

Hive does not have any support for splitting, but Pig has support for it. It even can apply different operators after splitting. Inserting a new query into an existing query is difficult in Apache Hive. In Pig Latin it is an easy thing to do it. You need to insert a new line of code and link the next line of code to the newly inserted relation.

Use Cases

Apache Pig can be used for every business case where Apache Hadoop is used. Here are some of them:

- Apache Pig is a widely used big data–processing technology. More than 60 percent of Hadoop jobs are Pig jobs at Yahoo.

- Twitter extensively uses Pig for log analysis, sentiment analysis, and mining of tweet data.

- PayPal uses Pig to analyze transaction data and fraud detection.

- Analysis of web logs is also done by many companies using Pig.

Pig Philosophy

Apache Pig has four founding principles that define the philosophy of Pig. These principles help users get a helicopter view of the technology. These principles also help developers to write new functionality with a purpose.

Pigs Eat Anything

Pig can process all types of data such as structured, semistructured, and unstructured data. It can also read data from multiple source systems like Hive, HBase, Cassandra, and so on. It supports many data formats such as text, sequence, ORC, and Parquet.

Pigs Live Anywhere

Apache Pig is big data–processing tool that was first implemented on Apache Hadoop. It can even process local file system data.

Pigs Are Domestic Animals

Like other domestic animals, pigs are friendly animals, and Apache Pig is user friendly. Apache Pig is easy to use and simple to learn. If a schema not specified, it takes the default schema. It applies the default load and store functions if not specified and applies the default delimiter if not given by the user. You can easily integrate Java, Python, and JavaScript code into Pig.

Pigs Fly

Apache Pig is used to build lightweight big data applications that have high performance. Apache Pig is instrumental in writing big functionality with few lines of code.

Summary

Traditional RDBMSs and data warehouse systems cannot scale to the growing size and needs of data management, so Google introduced two parallel computing technologies, GFS and MapReduce, to address big data problems related to storage and data processing.

Doug Cutting, inspired by the Google technologies, created a technology called Hadoop with two modules: HDFS and MapReduce. HDFS and MapReduce are the same as Google GFS and MapReduce with respect to functionality. MapReduce has proven itself in big data processing and has become the base platform for current and future big data technologies.

As MapReduce is low level and not developer friendly, many abstractions were created to address different needs. Cascading is a Java-based abstraction that hides the key and value from the end user and allows you to develop big data applications. It is not suitable for ad hoc querying. Apache Hive is data warehouse for big data and supports only structured data; even a nontechnical person can generate reports using BI tools on top of Hive.

Apache Pig is a scripting platform for big data processing that is developer friendly, is easy to learn, and can process all types of data.

CHAPTER 2

■ ■ ■

Data Types

In this chapter, you will start learning how to code using Pig Latin. This chapter covers data types, type casting among data types, identifiers, and finally some operators.

Apache Pig provides two data types. They are simple and complex, as specified in Figure 2-1. The simple data types include int, long, float, double, boolean, chararray, bytearray, datetime, biginteger, and bigdecimal. The complex data types are map, tuple, and bag.

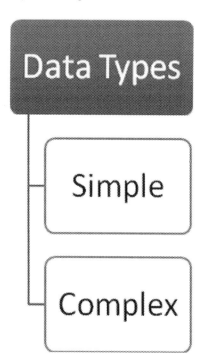

Figure 2-1. *Data types in Pig*

In Pig Latin, data types are specified as part of the schema after the as keyword and within brackets. The field name is specified first and then the data type. A colon is used to separate the column name and data type.

Here is an example that shows how to use data types:

```
emp = load '/data/employees' as (eid:int,ename:chararray,gender:chararray);
```

Now you will learn all about the data types in both the simple and complex categories.

Simple Data Types

In this section, you will learn about the simple data types in detail with example code, including the size they support. Most of data types are derived from the Java language.

int

The int data type is a 4-byte signed integer similar to the integer in Java. It can contain values in the range of 2^31 to (2^31)–1, in other words, a minimum value of 2,147,483,648 and a maximum value of 2,147,483,647 (inclusive).

The following shows some sample code that uses the int data type:

```
Sales = load '/data/sales' as (eid:int);
```

long

The long data type is an 8-byte signed integer that is the same with respect to size and usage as in Java. It can contain values in the range of 2^63 to (2^63)–1. A lowercase *l* or an uppercase *L* is used to represent a long data type in Java. Similarly, you can use lowercase *l* or uppercase *L* as part of constant.

The following is an example that uses the long data type:

```
emp = load '/data/employees' as (DOB:long);
```

float

The float data type is a 4-byte floating-point number that is the same as float in Java. A lowercase *f* or an uppercase *F* is used to represent floats, for example, 34.2f or 34.2F.

The following is an example that uses the float data type:

```
emp = load '/data/employees' as (salary:float);
```

double

double is an 8-byte floating-point number that is the same as double in Java. Unlike float or other data, no character is required to represent the double data type.

The following is an example that uses the double data type:

```
emp = load '/data/employees' as (salary:double);
```

chararray

The chararray data type is a character array in UTF-8 format. This data type is the same as string. You can use chararray when you are not sure of the type of data stored in a field. When you use an incorrect data type, Pig Latin returns a null value. It is a safe data type to use to avoid null values.

The following is an example that uses the chararray data type:

```
emp = load '/data/employees' as (country:chararray);
```

boolean

The boolean data type represents true or false values. This data type is case insensitive. Both True and tRuE are treated as similar in Boolean data.

The following is an example that uses the boolean data type.

```
emp = load '/data/employees' as (isWeekend:boolean);
```

bytearray

The bytearray data type is a default data type and stores data in BLOB format as a byte array. If you do not specify a data type, Pig Latin assigns the bytearray data type by default. Alternately, you can also specify the bytearray data type.

The following is an example that uses the bytearray data type:

```
emp = load '/data/employees' as (eid,ename,salary);
```

Here's another example:

```
emp = load '/data/employees' as (eid:bytearray,ename:bytearray,salary:bytearray);
```

datetime

The only date-based data type available in Pig Latin is datetime, which is used to represent the date and time. The data before *T* is the date, the data after the *T* is the time, the data after + is time zone.

23

The following is an instance of this format: 2015-01-01T00:00:00.000+00:00.

```
emp = load '/data/employees' as (dateofjoining:datetime);
```

biginteger

The biginteger data type is the same as the biginteger in Java. If you have data bigger than long, then you use biginteger. The biginteger data type is particularly useful for representing credit card or debit card numbers.

```
Sales = load '/data/sales' as (cardnumber:biginteger);
```

bigdecimal

The bigdecimal data type is the same as bigdecimal in Java. The bigdecimal data type is used for data bigger than double. Here's an example: 22.2222212145218886998.

Summary of Simple Data Types

Table 2-1 summarizes all the simple data types.

Table 2-1. *Simple Data Types*

Type	Description	Example
int	4-byte	100
long	8-byte	100L or 100l
float	4-byte	100.1f or 100.1F
double	8-byte	100.1e2
biginteger	Java biginteger	100000000000
bigdecimal	Java bigdecimal	100.0000000001
boolean	True/false	true/false
chararray	UTF-8 string	Big data is everywhere
bytearray	Binary data	Binary data
datetime	Date and time	2015:12:07T10:10:20.001+00.00

Complex Data Types

Complex data types in Pig Latin are used to process more than one data point. Complex data is classified as a map, tuple, or bag, as specified in Figure 2-2.

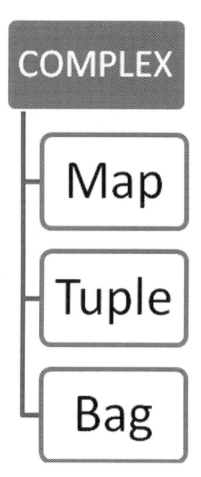

Figure 2-2. *Complex data types*

map

A map data type holds a set of key-value pairs. Maps are enclosed in straight brackets. The key and value are separated by the # character. The key should be the chararray data type and should be unique. While the value can hold any data type, the default is set to bytearray.

Here's the syntax (the key-value pair needs to be present in a file):

```
[key1#value1,key2#value2,key3#value3,...]
```

Here's an example:

```
[empname#Bala]
emp = load '/data/employees' as (M:map[]);
```

Here's another example:

```
emp = load '/data/employees' as (M:map[chararray]);
```

The second example states the value of the data type is chararray.

If your data is not in the map data type format, you can convert the two existing fields into the map data type using the TOMAP function.

The following code converts the employee name and year of joining a company to the map data type:

```
emp = load 'employees' as (empname:chararray, year:int);
empmap = foreach emp generate TOMAP(empname, year);
```

tuple

A *tuple* is an ordered set of fields and is enclosed in parentheses. A field can hold any data type including another tuple or bag. Fields are numbered beginning from 0. If field data is not available, then its value is set to a default of null.

For example, consider the following tuple:

```
(Bala,100,SSE)
```

This tuple has three data fields. Data in this example can be loaded using the following statement:

```
emp = load '/data/employees' as (T: tuple (empname:chararray,  dno:int,
desg:charray));
```

or the following statement:

```
emp = load '/data/employees' as (T:  (empname:chararray,  dno:int,
desg:charray));
```

You can convert existing data fields into tuples with the TOTUPLE function. The following code converts fields with simple data types into tuples:

```
emp = load '/data/employees' as (ename:chararray,  eid:int, desg:charray);
emptuple=foreach emp generate TOTUPLE(ename, eid,desg);
```

bag

A *bag* is a collection of tuples and is enclosed in curly brackets. A bag can have duplicate values. Tuples can have any number of fields. If the field value is not found, a null is returned.

Here is the syntax for the bag data type:

```
{tuple1, tuple2, tuple3,...}
```

Here's an example:

```
{(Bala, 1972, Software Engineer)}
```

Data for this example can be loaded using the following statement:

```
emp = load '/data/employees' as (B: bag {T: tuple (ename:chararray,
empid:int, desg:charray)} );
```

or the following:

```
emp = load '/data/employees' as (B: {T:  (ename:chararray,  empid:int,
desg:charray)});
```

There are two types of bags: outer bag and inner bag.
Here is an example of data with an inner bag:

```
(1,{( Bala, 1972, Software Engineer)})
```

You can convert fields with simple data types into bag data types using the TOBAG function.
The following lines of code convert existing fields into bag data types.

```
emp = load '/data/employees' as (ename:chararray,  empid:int, desg:charray);
empbag=foreach emp generate TOBAG(ename,empid,desg);

Dump empbag;

({(Bala),(1972),(Software Engineer)})
```

Summary of Complex Data Types

Table 2-2 summarizes the complex data types.

Table 2-2. *Complex Data Types*

Type	Description	Example
map	Key and value	[100#ApachePig]
tuple	Set of fields	(10,Pig)
bag	Collection of tuples	{(10,Pig), (11 Hive)}

Schema

Now you will look at how to check the available columns, and their data types, of a relation. To find out the schema of a relation, you can use the describe keyword.

The syntax for using describe is as follows:

```
Describe <<relation>>;
```

For example, the statement for describe usage is as follows:

```
emp = load '/data/emp' as (ename:chararray, eid:int, desg:chararray);
Describe emp;
```

The output will read as follows:

```
emp: {ename:chararray, eid:int, desg:chararray}
```

You need not specify the data type in Pig Latin, as it assigns the default data type bytearray to all columns.

The following code specifies column names without data types:

```
Movies = load '/data/movies' as (moviename, year, genre);
Describe Movies;
```

The output will read as follows:

```
Movies: {moviename:bytearray, year:bytearray, genre:bytearray}
```

Pig Latin allows you to load data without specifying a column name and data type. It uses the default numbering, starting from 0, and assigns the default data type. If you try to generate a schema for such a relation, Pig Latin will give the output Schema unknown. Whenever you perform operations on fields of such relations, Pig Latin will cast bytearray to the most suitable data type.

The following code does not specify column names and data types:

```
emp = load '/data/employees';
Describe emp;

emp: schema for emp unknown.
```

Casting

Casting is used to convert one data type to another. Pig Latin performs two types of casting: implicit casting occurs when Pig Latin performs casting automatically, and explicit casting occurs when a user performs casting. To perform explicit casting, you specify the target data type within parentheses.

To perform casting from int to chararray, you can use the following code:

empcode = foreach emp generate (chararray) empid;

bytearray is the most generic data type and can be cast to any data type. Casting from bytearray to other data types occurs implicitly depending on the specified operation. To perform arithmetic operations, bytearray casting is performed to double. Similarly, casting from bytearray to datetime, chararray, and boolean data types occurs when the user performs the respective operations. bytearray can also be cast to the complex data types of map, tuple, and bag.

Casting Error

Both implicit casting and explicit casting throw an error if they cannot perform casting.

For example, if you are performing a sum operation on two fields and one of them does not contain a numeric value, then implicit casting will throw an error.

If you are explicitly trying to cast from chararray to int and chararray does not have a numeric value, then explicit casting will throw an error.

Table 2-3 lists possible castings between different data types. For example, boolean can be cast to chararray but not to int because boolean is represented using true and false values, not 0 and 1.

Table 2-3. *Possible Castings Between Different Data Types*

From To	int	long	float	double	chararray	bytearray	boolean
int	NA	Yes	Yes	Yes	Yes	No	No
long	Yes	NA	Yes	Yes	Yes	No	No
float	Yes	Yes	NA	Yes	Yes	NO	No
double	Yes	Yes	Yes	NA	Yes	No	No
chararray	Yes	Yes	Yes	Yes	NA	No	Yes
bytearray	Yes	Yes	Yes	Yes	Yes	NA	Yes
boolean	No	No	No	No	Yes	No	NA

Comparison Operators

The operators in Table 2-4 are used in Pig Latin to perform comparison operations such as equal, not equal, greater than, and so on.

Table 2-4. Comparison Operations

Operator	Symbol
Less than	<
Greater than	>
Less than or equal	<=
Greater than or equal	>=
Equal	==
Not equal	!=
Pattern matching	matches

You can use the `filter` keyword to perform comparison operations. Here's an example:

```
Year= filter emp by (year==1972);
```

You can also perform pattern matching as part of the `filter` keyword using the `matches` operator.

For instance, the following code searches for employees who have `john` in their names:

```
empjohn = filter emp by (ename matches '.*john*.' ;
```

Identifiers

Identifiers may be names of fields, relations, aliases, variables, and so on. Identifiers always begin with an alphabet and can be followed by letters, numbers, and underscores.

Here are examples of correct names:

```
Moviename1_
Moviename5
```

Here are examples of incorrect names:

```
_movie
$movie
Movie$
```

Boolean Operators

Pig Latin includes Boolean operators such as AND, OR, NOT, and IN.

- AND returns true if all conditions are true.

- OR requires that at least one condition is true in order to return a true value.

- NOT inverts the value, and IN represents a list of "or" conditions.

Boolean operators are specified in the filter statement.

The following code filters employees whose year of joining a company is either 1997 or 1980:

```
empyoj = FILTER emp BY (yearOJ==1997) OR (yearOJ==1980);
```

Summary

In this chapter, you learned the fundamentals of Pig Latin. Simple data types, complex data types, and type casting among them were discussed. You also learned how to write an identifier and how to use comparison operators and Boolean operators.

CHAPTER 3

Grunt

In the previous chapter, you learned the Pig Latin fundamentals such as data types, type casting among data types, and operators. In this chapter, you will learn about the command-line interface (CLI) of Pig, called Grunt.

You can use Grunt for submitting Pig Latin scripts, controlling jobs, and accessing file systems, both local and HDFS, in Pig.

Invoking the Grunt Shell

You can start a Grunt shell in the following four modes: Local, MapReduce, Tez, and Tez-local.

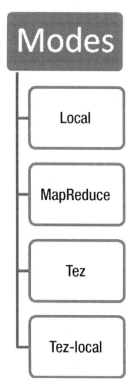

Figure 3-1. *Grunt modes*

© Balaswamy Vaddeman 2016
B. Vaddeman, *Beginning Apache Pig*, DOI 10.1007/978-1-4842-2337-6_3

MapReduce is the default invoke mode. Type **pig** to start the Grunt shell in MapReduce mode. However, if exectype is specified in the `pig.properties` file, then exectype becomes the default mode.

Local mode reads data from the local file system and runs MapReduce jobs. Local modes are used to process small amounts of data on a single machine.

Pig version 0.14 has a new execution engine called Tez that incorporates Apache Tez as an execution engine. Tez is a framework developed on top of YARN to address performance issues involving input/output (I/O) in MapReduce. Tez jobs will be much faster than MapReduce. You will learn more about Tez in Chapter 17.

Tez-local also reads data from the local file system but using the Tez framework.

You'll now see some examples for starting the Grunt shell.

Here's the syntax:

```
Pig  -x mode
```

Here are some examples:

```
Pig -x tez
Grunt>

Pig -x local
Grunt>

//default mode
Pig
Grunt>
```

The -x option is the short form of -execctype.

Work is in progress to add Spark as an execution engine. Once it is available, you can start the Spark execution engine with the `pig -x spark` command.

Now you will learn about the various commands available in the Grunt shell.

Commands

Two commands in Pig Latin—namely, fs and sh—help you interact with file systems and run shell scripts.

The fs Command

The fs command is used to invoke the fsshell commands of HDFS. This command can be executed in the Grunt shell and Pig Latin scripts.

Here are a few examples of the sh command to check the input file path in the Grunt shell before specifying it as an input path and to check the output directory once the job is completed. Without the sh command, you would have to exit Grunt to check the input file path and return to Grunt to continue. Thus, the command saves time.

The syntax for the fs command follows:

```
Fs <fs-command> <input-parameters>

fs-command        : fsshell-command like ls,mkdir or rm etc.

input-parameters  : parameters of above command.
```

For example, you can use the following fs command to create a directory:

```
grunt >fs -mkdir /data
```

You ca use the following the fs command to list the directory or file:

```
grunt >fs -ls /data
```

You can refer the file system shell guide for a complete list of fsshell commands (http://xxx).

The fs command chooses the relevant file system to interact with, based on the mode in which you have chosen to invoke Pig. If you have started Pig in Local mode, the fs command will interact with the local file system; otherwise, it will interact with the Hadoop distributed file system. Some commands such as ls will run without input parameters and will take the current working directory in Local mode and the user home directory in nonlocal mode.

The sh Command

sh is used to run shell commands such as ls, mkdir, and du from both Grunt and Pig Latin scripts. However, commands such as cd are not relevant with the sh command.

Use the following syntax to execute sh commands:

```
Sh <shell-command> <input-parameters>

Shell-command      : operating system commands like ls, mkdir and rm etc.
input-parameters   : parameters of above command.
```

The following are examples of sh commands:

```
grunt> sh ls /
wordcount.pig
joinsdemo.pig

grunt> sh mkdir /data
```

If Pig is started in Local mode, both the fs and sh commands might have the same effect. In Local mode, if you create a directory called test using either the fs or sh command, a folder called test is created on the local file system as well.

You have learned how to use the fs and sh commands to interact with both the local file system and the Hadoop distributed file system. Now you will learn about other utility commands that are useful for controlling both the Grunt shell and Hadoop jobs.

Utility Commands

The following are the utility commands.

help

The help command prints a list of Pig commands or properties with a short description and syntax.

```
grunt> help

Commands:
<pig latin statement>; - See the PigLatin manual for details: http://hadoop.
apache.org/pig
File system commands:
    fs <fs arguments> - Equivalent to Hadoop dfs command: http://hadoop.
    apache.org/common/docs/current/hdfs_shell.htm
```

You can display properties with the properties option, as in the following example:

```
Pig  -help properties
```

history

The history command displays previously used commands in the Grunt shell. You can also rerun them. The commands are numbered by default. To remove the numbering, use the -n option. To navigate the list of commands, use the up arrow and the down arrow.

The following are the history command examples:

```
Grunt>history
1   num = load 'movies';
2   num11 = filter num by $0==11;

Grunt>history -n
num = load 'movies';
num11 = filter num by $0==11;
```

quit

Use the quit command to exit the Grunt shell.

```
Grunt>quit;
hdfs@cluster5en1:~>
```

kill

The kill command specifies the job ID to abort a running job. The kill command is equivalent to the mapred job -kill <jobid> command in Hadoop. Most of the time, you will not remember job IDs, so you can use the mapred job -list command to get currently running jobs.

```
Grunt> kill job_201512120001_001;
```

set

The set command applies a value to a property name, for both Pig and Hadoop. It can be used in the Grunt shell and Pig Latin scripts. This section shows Pig properties for which you can set values.

The set default_parallel command specifies the default number of reducers; in this example, it sets the default number of reducers to 20:

```
Grunt>set default_parallel 20;
```

The set debug command enables and disables debugging in a Pig Latin script. It is set to disable, by default. The following command enables debugging:

```
Grunt>set debug 'on';
```

To then disable it, use the off option as follows:

```
Grunt> set debug off;
```

This set command allows you to set a name for a job. By default PigLatin:scriptname is given as the job name if the job is submitted from the Grunt shell.

```
PigLatin: DefaultJobName
```

The following example sets the job name to test:

```
Grunt> Set job.name 'test';
```

The set job.priority command allows you to set a priority for a job. It can take the value very_low, low, normal, high, or very_high. The default priority is normal. The following example sets the priority to low:

```
Grunt>  set job.priority low;
```

This command is also used to set values to Hadoop properties. The following line sets the number of reducers to 5, using the Hadoop property mapred.reduce.tasks:

```
Grunt> set mapred.reduce.tasks 5;
```

clear

This clears all visible commands in the Grunt shell:

```
grunt>clear
```

exec

The exec command runs Pig Latin scripts from the Grunt shell.
    Here's the syntax:

```
Exec [-param ] [-param file ] pigscript
```

- -param: This specifies extra parameters such as a name-value pair and is optional.

- -paramfile: Lengthy scripts often have many dynamic parameters defined. While it is difficult to read a lengthy command, it is also difficult to maintain values provided at the time of running the script. The -paramfile option helps manage lengthy scripts by creating a file containing all the property names and their values.

The command looks like the following:

```
Grunt>cat /home/hdfs/allproperties.props;
Inpath=/user/hdfs/data
```

```
Grunt> exec –paramfile /home/hdfs/allproperties.props  /home/hdfs/
dumpmovies.pig;
```

If there are multiple parameters, you can specify all the property names and their values in a file. The file path can also be specified to a Pig script. This is optional.

```
Grunt> exec dumpmovies.pig;
```

You can specify both the absolute and relative paths. The relative path will be resolved from the current working directory of the local file system.
    The -param option allows you to provide values for dynamic parameters defined in the script. You can define dynamic properties in a script using the $ symbol.

The following code defines a dynamic parameter called `inpath`:

```
Movies = load '$inpath';
Dump movies;
```

The value for `inpath` can be specified at the time of running the script using the
`-param` option, as shown here:

```
Grunt> exec –param inpath=/home/hdfs/movies dumpmovies.pig
```

run

The `run` command is similar to the exec functionality, and it executes a Pig Latin script
from Grunt. But it also has some additional features. One feature is that commands used
in the script are saved in history. Thus, they can be executed using aliases from the script
after running it.

Here's the syntax:

```
run [-param] [-param_file] piglatinscript
options are same as in command exec.
```

- `-param`: This specifies extra parameters such as a name-value
  pair. It is not mandatory because some scripts may not have
  parameters.

- `-paramfile`: This specifies all property names and their values in
  a file when handling multiple parameters. Some scripts may not
  have params, so this is optional.

Here's an example:

```
Grunt> run –param inpath=/home/hdfs/movies  dumpmovies.pig
```

Summary of Commands

Table 3-1 describes the various commands and shows an example for quick reference.

Table 3-1. *Commands*

Type	Command	Short Description	Example
File system	fs	File system commands	Grunt>fs -ls /
Shell	sh	Runs shell programs	Grunt>sh ls /
Utility	exec	Runs Pig Latin scripts from the Grunt shell	Grunt>exec dumpmovies. pig
	run	Runs Pig Latin scripts from the Grunt shell	Grunt>run dumpmovies. pig
	clear	Clears all commands from the Grunt shell	Grunt>clear
	help	Displays all command information	Grunt>help
	history	Displays previously run statements	Grunt> history
	set	Sets a value to a property	Set debug 'on'
	quit	Quits the Grunt shell	Grunt>quit;
	kill	Kills a job	Grunt>kill job_201512120001_001

Auto-completion

Since the Grunt shell is equipped with the auto-completion feature, you do not need to type out complete commands. Type the first few characters of a command and press the Tab key; the rest of the command is inserted automatically. If more than one command begins with the specified characters, all of them are displayed. For example, if you press L and press the Tab key, you will see load, long, and ls.

```
grunt> movies = l --hit tab

load    long    ls

grunt> movies = l
```

Summary

In this chapter, you learned the execution modes available for Pig Latin and also learned how to interact with both the local file system and HDFS using the fs and sh commands. You also gained knowledge of how to kill Hadoop jobs using the kill command and how to control the Grunt shell using the help, history, exec, and run commands.

CHAPTER 4

Pig Latin Fundamentals

In this chapter, you will learn the basics of Pig Latin. You will learn how to run Pig Latin code, and you will come to understand Pig Latin basic relational operators and parameter substitution.

Running Pig Latin Code

You can run Pig Latin code in several ways:

- Using the Grunt shell
- Using the `pig -e` command
- Using the `pig -f` command
- Using the Hue tool
- Embedding Pig Latin code in Java code

Let's see how to run the code using each of these methods.

Grunt Shell

As discussed in Chapter 3, you can use the Grunt shell to submit Pig Latin code. Since code submitted from the Grunt shell is session-based, once you exit the Grunt shell, you lose the code, and you have to start fresh from the first line of code. The Grunt shell is useful to run small Pig Latin code for testing in development and testing environments.

Here's some sample code in the Grunt shell:

```
Grunt>emp = load '/data/employee' using PigStorage() as (eno:int,ename:chara
rray,salary:chararray,deptno:int);
Grunt>
```

Pig -e

When you have to run a list of Pig Latin statements, you can use the `pig -e` command. Here's the syntax:

```
Pig -e " <pig latin statement>"
```

Although Pig Latin code is not easy to read, it is easy to modify. This command can also be embedded easily in the shell script for further automation.

The following code uses the `Pig -e` command, and the semicolon represents the end of a statement:

```
Pig -e "emp = load '/data/employee' using PigStorage(',');dump emp;"
```

Pig -f

The `Pig -e` command is not readable if it has multiple lines of code. `Pig -f` allows you to embed Pig Latin code in a file. Here's the syntax:

```
Pig -f /path/to/piglatin/file
```

Although the file can be saved with any extension, it is better to save the file with a `.pig` extension to distinguish it from other script files.

This is the standard way to write Pig Latin code in a production environment. Running Pig Latin code this way helps you write more dynamic scripts as it supports dynamic parameters, allows you to write reusable Pig Latin scripts, and helps maintain the scripts easily.

The following is an example of the `pig -f` command:

```
Cat /home/hdfs/dumpemp.pig
emp = load '/data/employee' using PigStorage(',')  as (eno:int,ename:chararr
ay,salary:int,deptno:int) ;
dump emp;

pig -f /home/hdfs/dumpemp.pig
```

Embed Pig Code in a Java Program

You can also run Pig Latin code using Java. The `PigServer` class in Java interacts with the Pig Latin code. As discussed in the previous chapter, you can specify the execution mode, either MapReduce, Tez, Local, or Tez-local, in the `PigServer` constructor. If you choose Local mode, Pig runs the job on the local file system, and MapReduce runs the job on the distributed file system.

The following code instantiates the `PigServer` class:

```
PigServer ps=new PigServer("local");
```

The following are some useful methods of the PigServer class:

- Use the registerQuery() method of the PigServer class to register Pig Latin code. However, code specified in registerQuery is not run until you submit the store or dump command.

- Use the registerScript() method of the PigServer class to register a file as a Pig Latin script file.

- Use the store() method of the PigServer class to start the execution of Pig Latin code. It takes two inputs: the relation name and the output directory. It stores the given relation name results in the specified output directory.

- Use the registerJar() method of the PigServer method to register a JAR file if it contains any user-defined functions.

The following Java code reads data from the input directory and writes it to the output directory. However, observe that the input and output directories are not specified in the Java code; hence, you would have to provide them at runtime.

1. Write the following Java program:

```
import java.io.IOException;
import org.apache.pig.PigRunner;
import org.apache.pig.PigServer;
import org.apache.pig.backend.executionengine.ExecException;

public class StoreEmp {
   public static void main(String[] args) {

      PigServer pigServer=null;
      try {
         pigServer = new PigServer("local");
         pigServer.registerQuery("emp = load '"+args[0]+"' ;");
         pigServer.store("emp", args[1]);

      } catch (ExecException e1) {
         e1.printStackTrace();
      } catch (IOException e1) {
         e1.printStackTrace();
      }
      }
      }
```

If you choose to write this program using Eclipse, make sure you have the pig.jar file in the build path to check and compile it successfully. Copy the file to one of the cluster nodes to run the program on a cluster.

2. Write the command to compile the Java program.

   Generate the .class file for the .java file by compiling it with the javac program. Mention Pig*.jar in the classpath so that the Pig API in the Java program is resolved.

   The following command compiles Java file.

   ```
   /usr/lib/jdk8/bin/javac -cp /usr/hdp/2.3.4.0-3485/
   pig/pig-0.15.0.2.3.4.0-3485-core-h2.jar StoreEmp.
   java.
   ```

3. Write the command to run the Java program.

   After generating .class for the .java file, run it using the Java program shown here:

   ```
   /usr/lib/jdk8/bin/java -cp /usr/hdp/2.3.4.0-3485/pig/pig-
   0.15.0.2.3.4.0-3485-core-h2.jar:/usr/hdp/2.3.4.0-3485/
   hadoop-hdfs/lib/*:/usr/hdp/current/hadoop-client/client/*:/usr/
   hdp/2.3.4.0-3485/pig/lib/*:. StoreEmp employee.csv dumpempout
   ```

If Pig cannot find its dependent JARs, the Java program might fail and throw a "class not found" exception. To avoid such exceptions, include all the required JARs in the class path using the -cp option.

The following are other special characters that can be used after the –cp option.

- *Colon (:)*: When multiple directories are required, the colon delimits the directories.

- *Asterisk (*)*: This includes all files from a given directory in the class path.

- *Dot (.)*: This adds the current directory to the class path mentioned, so as to take the generated .class file from the current directory.

After specifying the class name StoreEmp, you specify the input and output directories; you would not have specified them in Java code. In the previously mentioned code, employee.csv is the input directory, and dumpempout is the output directory.

Hue

Hue contains a list of web applications that allows users to submit Hadoop tools code to a cluster. It even has a Pig Editor to submit Pig Latin code. You will learn more about it in Chapter 8.

Pig Operators and Commands

I will now discuss some operators and commands available in Pig. Pig operators are the pillars of Apache Pig and are used for data analysis.

Pig operators and commands are categorized into Reduce operators and non-Reduce operators, as shown in Figure 4-1. Since most tasks are resource-intensive, Reduce operators such as join, cogroup, and distinct launch the Reduce tasks. The reduce operators lets you stipulate the number of Reduce tasks that you want to run at a given time for optimal performance. Other operators such as limit and flatten do not launch Reduce tasks.

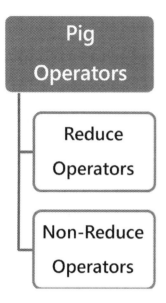

Figure 4-1. *Pig operators*

Now you will learn about both Reduce and non-Reduce operators in detail.

Load

The load operator loads data from the source system. Its syntax is followed by two operators: as and using.

An example of load follows:

```
emp = load '/data/employee' using PigStorage(',') as (eno:int,ename:chararra
y,salary:int,deptno:int);
```

The as operator defines the schema, and using specifies the function that you applied while reading data. By default, Pig Latin chooses PigStorage() for both the schema and the function. bytearray is the default data type in the default schema. The number list starting at 0 is taken as the default field name. Pigstorage('\t') applies a tab as the default delimiter, but you can specify any other character as the delimiter.

load allows you to read from both directories and files. You can also specify the absolute and relative paths of files and directories. The relative path is resolved with respect to the current working directory in Local mode. The user home directory in HDFS is the default consideration for the relative path in MapReduce and Tez modes. You can also specify multiple paths with a comma separator. Compressed files are automatically loaded using the corresponding codecs in Pig Latin. If you have .gz files as the source data sets, you need not direct the codec to read the data. The load operator reads it automatically.

You can even use regular expressions (RegEx) in the file path of the load statement.

RegEx in the File Path

Sometimes you may have to use multiple files and directories in the load statement. RegEx characters help specify multiple directories and files.

Here are some RegEx characters that you can use in load statements:

SNO	RegEx Characters	Meaning
1	?	Resolves to any single character
2	*	Resolves to zero or more characters
3	[abc]	Resolves to a single character from the character set {a,b,c}
4	[a-b]	Resolves to a single character from the character range {a...b}
5	[^a]	Resolves to a single character that is not from the character set or range {a}
6	\c	Removes any special meaning of character c
7	{ab,cd}	Resolves to a string from the string set {ab, cd}
8	{ab,c{de,fh}}	Resolves to a string from the string set {ab, cde, cfh}

These RegEx characters are taken from HDFS GLOB file patterns. The following file paths will resolve to all employee.csv files whose parent directories range from date=2016-07-01 to date=2016-07-07.

The following HDFS example code lists all files:

```
/user/hdfs/date=2016-07-0[1-7]
/user/hdfs/date=2016-07-0[1234567]
```

To test a RegEx string using the HDFS shell guide, use the following code:

```
hdfs@cluster10-1:~> hdfs dfs -ls /user/hdfs/date=2016-07-0[1-7]
Found 1 items
-rw-r--r--   3 hdfs hdfs          85 2016-07-04 05:27 /user/hdfs/date=2016-
                                                      07-01/employee.csv
```

```
Found 1 items
-rw-r--r--   3 hdfs hdfs      85 2016-07-04 05:27 /user/hdfs/date=2016-
                                  07-02/employee.csv
Found 1 items
-rw-r--r--   3 hdfs hdfs      85 2016-07-04 05:27 /user/hdfs/date=2016-
                                  07-03/employee.csv
Found 1 items
-rw-r--r--   3 hdfs hdfs      85 2016-07-04 05:27 /user/hdfs/date=2016-
                                  07-04/employee.csv
Found 1 items
-rw-r--r--   3 hdfs hdfs      85 2016-07-04 05:27 /user/hdfs/date=2016-
                                  07-05/employee.csv
Found 1 items
-rw-r--r--   3 hdfs hdfs      85 2016-07-04 05:27 /user/hdfs/date=2016-
                                  07-06/employee.csv
Found 1 items
-rw-r--r--   3 hdfs hdfs      85 2016-07-04 05:27 /user/hdfs/date=2016-
                                  07-07/employee.csv
```

Rather than specifying all the file paths, with RegEx you can specify a single file path that gets resolved to all the files.

The following code contains a load operator with a RegEx string:

```
emp = load '/user/hdfs/date=2016-07-0[1-7]/employee' using
    PigStorage(',') as (eno,ename,salary,dno);
```

or as follows:

```
emp = load '/user/hdfs/date=2016-07-0[1234567]/employee' using
    PigStorage(',') as (eno,ename,salary,dno);
```

store

store writes a relation data to a sink or output directory. The sink can be an HDFS, HBase, or Accumulo. The store statement is followed by the INTO and USING keywords.

The INTO keyword is mandatory in the store operator's syntax, and it specifies the sink. store can also contain the file system relative path that is resolved based on the selected execution engine. If Pig Latin is working in Local mode, the relative path is resolved with respect to the current working directory, and if it is nonlocal mode, it is resolved to the user home directory in HDFS.

using specifies a store function, and PigStorage is the default function. Pig Latin code fails if an output directory already exists. With store, you can write data in different formats such as ORC, JSON, and Avro. You can also write data to different data stores such as HBase and Accumulo. You will learn in detail how to write different formats and how to write data to different data stores in Chapter 6.

The following code writes data of the relation deptcount to the output directory /data/deptcount using a comma delimiter:

```
Store deptcount into '/data/deptcount' using PigStorage (',');
```

> **Write a Pig Latin script that changes the default delimiter in the source directory (/data/employee) to a comma and stores data to a different directory.**

dump

The command dump displays data of a relation on the console. Write the relation name after the dump operator to display relation data on the console.

```
Dump <relationname>
```

With dump you can test code prior to production. You can also use it to debug Pig Latin scripts. Any Pig Latin code that follows the dump operator is ignored and is not executed.

Developers can insert a dump line at any step in the Pig Latin code to view input data at that particular step. All the data is displayed on the screen. So, if a relation has large data, it will take so much time to display the data on the console.

The following code displays five records from the employee data set on the console:

```
Emp = load '/data/employee' using PigStorage(',') as (eno:int,ename:chararray,
     salary:int,deptno:int);
Records5 = limit emp 5;
Dump records5;
```

version

Use the version option to check the version of Pig you are using.

```
Pig --version
```

Foreach Generate

Foreach is a multipurpose operator used for projection, applying functions, generating a new schema, and performing nested operations. Often, it is used along with the generate keyword, and it operates on one row at a time from the specified relation. Most of its functionality is similar to a SQL SELECT clause. Foreach, also called the transformation operator, performs transformation jobs.

Here's the syntax:

```
relname  = FOREACH relname1 { block | nested_block };

relname1    : Relation name to be used
```

block is used to process outer bags. You can use the foreach .. generate statement to process data. Nested_block is used to process inner bags.

Projection

You can choose or project some or all fields from a relation using the foreach .. generate statement. The following code chooses all the fields from the emp relation:

```
All = foreach emp generate *
```

The asterisk represents all the fields of a relation. You may have to use the relation name to access a field if you have bag, tuple, or map data types involved, as shown here:

```
Empnoname= foreach emp generate emptuple.empno,emptuple.empname;
```

Flatten

Apply the FLATTEN operator using the foreach .. generate statement to change structure of data from a bag to a tuple and from a tuple to a field. For more information, read about the Flatten operator later in the chapter.

Using Functions

The foreach .. generate statement is also used to apply a function on a field or set of fields. Functions include the following:

- Built-in

- User-defined

- Single-row functions such as UPPER

- Multirow functions such as COUNT, which must be applied after the GROUP operator

The following foreach statement converts all employee names into uppercase:

```
Enameucase = foreach emp generate UPPER(ename);
```

You will learn how to use the other functions such as lower, max, and count in Chapter 5.

New Schema

You can define a new schema on the output of the foreach .. generate statement.
The following code increases the salary by 10 percent and changes its data type:

```
emp = load 'employee.csv' using PigStorage(',') as (eno:int,ename:chararray,
    salary:int,dno:int);
describe emp;
emp: {eno: int,ename: chararray,salary: int,dno: int}
```

```
Newsal = foreach emp generate eno,ename,salary*1.1,deptno as
        (eno :int,ename:chararray,salary:doouble,deptno:int)
Describe Newsalary;
newsal: {eno: int,ename: chararray,newsal: double,dno: int}
```

Nested Block

Nested blocks process inner bags. Multiple operations can be performed within nested blocks. Nested blocks are enclosed in opening ({) and closing brackets (}). The last statement should always be generate. Macros cannot be used in nested blocks.

The following code displays the employee name and department name by retrieving them from different inner bags after the cogroup operator:

```
emp = load 'employee.csv' using PigStorage(',') as (eno:int,ename:chararray,
      salary:int,dno:int);
dept = load 'dept.csv'using PigStorage(',') as (dno:int,dname:chararray);
cogrp = cogroup emp by dno,dept by dno;
describe cogrp;
```

Here is a sample of the cogrp schema:

```
cogrp: {group: int,emp: {(eno: int,ename: chararray,salary: int,dno: int)},
dept: {(dno: int,dname: chararray)}}
```

You can process inner bags using nested blocks, as shown here:

```
enamedname = foreach cogrp{
 generate flatten(emp.ename),flatten(dept.dname);
}
```

filter

filter retrieves data based on specified conditions. filter checks one row after another for the specified conditions.

It is best practice to filter data as early as possible in the project so that you will get an early opportunity work on less data.

The following are some conditions that can be applied using the filter operator.

null

As discussed in Chapter 2, you can check for null values in a row using the null operator. The following code retrieves employee records in which deptno is null.

```
Nodept=filter emp by deptno is null:
```

Boolean Operators

You can apply AND, OR, and NOT Boolean operators in a filter statement.

The following code retrieves employee records whose deptno is greater than or equal to 100 and less than or equal to 300:

```
somedept=filter emp by deptno>=100 and deptno<=300;
```

Comparison Operators

You can apply the comparison operator using the filter statement.

The following code retrieves all employee records whose deptno is 300:

```
dept300=filter emp by deptno==300;
```

Limit

limit retrieves a specific number of tuples from a relation. limit is followed by the relation name and number.

```
Limit relationname number
```

The following code retrieves only five tuples from the relation emp. If five tuples are not available, it returns only the available number of tuples, and if more than five tuples are available, only five tuples are retrieved. Tuples are returned randomly, and the same tuples may not be retrieved when you rerun the code until and unless you apply the order by operator, before the limit operator.

The following code is a limit example:

```
Limit5 = limit emp 5;
```

This operator is useful to check data in a relation in the initial stage of writing Pig Latin code.

Assert

Assert verifies a specified condition on a relation. If a condition is not true, it throws the appropriate error message.

Here's the syntax:

```
ASSERT relname BY condition [, errormessage];
```

The following code verifies deptno is not null in the emp relation. If deptno is null, the MapReduce job will fail with the assertion violated error deptno should not be null.

```
Emp = load '/data/employee' using PigStorage() as (eno:int,ename:chararray,
      salary:int,deptno:int);
Assert emp by deptno is not null,'deptno should not be null';
Dump emp;
```

This operator is also useful for testing data for abnormalities in the development and testing environments.

You need to be sure when using it in production for a couple of reasons.

- By design, it fails the MapReduce job if the condition evaluates to true.

- It also causes performance issues.

SPLIT

split divides the data of a relation into two or more relations based on the condition you specify. It is opposite of the join functionality, which retrieves data from two or more relations into one relation.

Here's the syntax:

```
SPLIT relationname INTO
              relname1 IF condition1,
              relname2 IF condition2,
              ... ,
              relname3 OTHERWISE;
```

Data that fulfils the specified condition is moved to the specified relation.

```
relname1 IF condition1
```

Data that does not fulfil any of the mentioned conditions is moved to the default relation specified with otherwise. The otherwise statement is optional.

```
relname3 otherwise
```

The following code splits the emp relation data into three relations. Relation dept200 contains the employee records whose department number is less than or equal to 200. Relation dept300 contains the employee records whose department number is greater than 200 and less than or equal to 300. Employee records that do not satisfy the previous conditions will go into the otherdept relation. These three relations data is written to the output directory using the store command.

```
split emp into
dept200 IF deptno<=200 ,
dept300 IF (deptno>200 and deptno<300),
otherdept otherwise;
```

```
store dept200 into 'dept200';
store dept300 into 'dept300';
store otherdept into 'otherdept';
```

The split functionality is same as the filter functionality if you do not process result relations or store them in some output directories.

SAMPLE

To build statistical models, for instance, you do not need complete data from a relation. Under such circumstances, you can use the sample operator to generate sample data.

```
Sample relname  expression.
```

expression can be a number between 0 and 1. That is, .2 means 20 percent. You can use a scalar like 100/200. The following code generates 10 percent random data from the relation emp.

```
emp10 = Sample emp .1
```

The same tuples may not be displayed if you rerun the code.

Many algorithms are available to build sample data, and Pig Latin does not provide rich sampling techniques.

For further sampling algorithms such as Bernoulli sampling and Wiegthed random sampling, you can use the DataFu project. See https://datafu.incubator.apache.org/docs/datafu/guide/sampling.html.

FLATTEN

Most Pig Latin operators generate output in complex data types like bags or tuples. How do you process such data? You can use the FLATTEN operator to change the structure of the bag and tuple data types. When FLATTEN is applied, it changes the tuple structure to a field structure and the bag structure to a tuple.

Tuple Example

Before applying flatten, the tuple structure appears thusly:

```
emp: {et: (eno: int,eame: chararray,salary: int,deptno: int)}
```

The following code applies flatten on the tuple structure and changes it to a field structure:

```
emp = load 'emptuple.csv' using PigStorage(',') as (et:tuple(eno:int,eame:ch
ararray,salary:int,deptno:int));
describe emp;
```

If you apply flatten on a tuple as specified in the previous code, the structure of the data and schema will change to a field, as shown here:

```
emp = foreach emp generate FLATTEN(et);
describe emp;

emp: {et::eno: int,et::eame: chararray,et::salary: int,et::deptno: int}
```

Bag Example

A bag structure changes to a tuple structure when flatten is applied. Sometimes a cross join is performed when you apply the flatten operator. flatten also removes empty bags.

The following code performs a group operation on the deptno field of the emp relation and selects the group field and ename field, which are not actually flattened:

```
deptgrp = GROUP emp by dno ;
empcount = foreach deptgrp generate group,emp.ename;
dump  empcount;
```

Without applying flatten, data will have three tuples and appear like so:

```
(200,{(Niruopam),(Bala)})
(300,{(Radha)})
(,{(Nitya)})
```

If you apply flatten on emp.ename, a cross-join is performed, and four tuples are produced.

```
empcount = foreach deptgrp generate group,flatten(emp.ename);
dump  empcount;

(200,Niruopam)
(200,Bala)
(300,Radha)
(,Nitya)
```

import

The import operator is used to import a Pig Latin macro. You will learn about it in Chapter 10.

define

define is used to define a macro and an alias for lengthy and complex Pig Latin commands. You will learn how to use define in macros in Chapter 10 and define in user-defined functions in Chapter 11.

distinct

The distinct operator removes duplicate tuples from a relation. Its functionality is the same as a SQL distinct clause. The SQL while distinct can be applied on a field; the Pig Latin can be applied only on a relation.

For better results, you need to apply order by before the distinct operator. Here's the syntax:

```
Distinct relname [PARTITION by partitionercalss] [PARALLEL num]
```

The following code removes duplicate records from the emp relation:

```
Emp = load '/data/employee' using PigStorage(',') as (eno:int,ename:chararray,
    salary:int,deptno:int);
uniqemp = distinct emp;
dump uniqemp;
```

Choosing the Number of Reduce Tasks

The distinct operator launches the Reduce task. You can specify the number of reducers after the PARALLEL keyword.

The following code launches ten Reduce tasks to remove duplicate employee tuples:

```
uniqemp = distinct emp PARALLEL 10;
```

Using the MapReduce Partitioner

The MapReduce partitioner decides on the Reduce tasks for map output depending on a key. You can use the MapReduce partitioner in the distinct operator.

The following code applies HashPartitioner while performing the distinct operation:

```
numcount = distinct num partition by org.apache.hadoop.mapreduce.lib.
partition.HashPartitioner;
```

RANK

The rank operator provides a rank to every tuple of the specified relation. By default, rank would start ranking tuples from 1. rank can also be assigned after ordering one or more fields.

Here's the syntax:

```
Relname2 = RANK relname1 [By Fieldname1 ASC|DESC, By Fieldname2
ASC|DESC,...] [DENSE]
```

The following code shows an example of the rank operator.

```
emprank = RANK emp;

(1,100,Bala,100000,200)
(2,200,Radha,200000,300)
(3,300,Nitya,150000,)
(4,400,Niruopam,1600000,200)

emprank = RANK emp by empno DESC;
```

The previous code applies the rank after ordering empno in descending order. By default, if there is a tie for some tuples, the same rank is assigned, and the next tuple will not get the next rank. In the following data, Nitya gets tuple 2, and 2 ranks, so 3 is not assigned to any tuple. The next rank starts from 4.

```
(1,400,Nirupam,1600000,200)
(2,300,Nitya,150000,)
(2,300,Nitya,150000,)
(4,200,Radha,200000,300)
(5,100,Bala,100000,200)
```

To generate rank consecutively, you must apply a dense rank.
The following code shows an example of dense ranking:

```
emprank = RANK emp by empno DESC DENSE;

(1,400,Nirupam,1600000,200)
(2,300,Nitya,150000,)
(2,300,Nitya,150000,)
(3,200,Radha,200000,300)
(4,100,Bala,100000,200)
```

Union

The union operator is used to merge two or more relations.

```
Relname3 = UNION [ONSCHEMA] relname1, relname 2, ...] [PARALLEL n];

ONSCHEMA            -schema is mandatory for a relation if it is specified.
Relname1,relname2,..    -relations to be merged
PARALLEL n             -decides number of output files
```

The union output might contain duplicate tuples because the union operator does not remove them.

The following code shows an example of union:

```
Emp1= load '/data/employee' using PigStorage(',') (eno:int,ename:chararray,
      salary:int,dno:int);;
Dump emp1;
(200,Radha,200000,300)
(400,Nirupam,1600000,200)

Emp2 = load '/data/newemployee' using PigStorage(',')  (eno:int,ename:chararray,
      salary:int,dno:int);
Dump emp2;

(400,Nirupam,1600000,200)
(100,Bala,100000,200)
(300,Nitya,150000,)
Empuni = union emp1,emp2;
Dump empuni;

(200,Radha,200000,300)
(400,Nirupam,1600000,200)
(100,Bala,100000,200)
(400,Nirupam,1600000,200)
(300,Nitya,150000,)
```

The union functionality is achieved only by using the Map task. It does not launch any Reduce tasks. But you can use the parallel keyword to get n number of output files.

The parallel keyword with union works only in the Tez execution mode.

ONSCHEMA checks for the schema of all relations used.

The union operation fails even if any one of the relations does not have the schema defined.

The following code fails because emp1 does not have any schema defined:

```
Emp1 = load '/data/employee' using PigStorage(',') ;
emp2 = load '/data/newemployee' using PigStorage(',') as (eno:int,ename:char
array,salary:int,dno:int);
empuni = union emp1,emp2 parallel 10;
```

In the union output, the order of tuples is not preserved. Pig Latin allows relations of different schemas to be merged. You must be cautious while processing the union output because a union between relations with different schemas and sizes can cause changes in the schema of the union output.

ORDER BY

The Order By operator works the same way as the SQL ORDER BY. It sorts relation data using the mentioned fields. The order can be ascending or descending.

Here's the syntax:

```
relname = ORDER relname1 BY [ *|fieldnames [ASC|DESC]] [PARALLEL n];
*                -used to specify complete tuple.
Fieldnames       -one or more fileds can be specified. order.
Asc              -sorts a field in ascending order.
Desc             -sorts a field in descending order.
Parallel n       -used to specify n number of reduce tasks.
```

The following code orders the emp relation on the department number field:

```
Deptorder = ORDER emp by deptno;
```

Specifying an order is optional, and by default ascending order is applied. Null values are returned first in ascending order and last in descending order. Pig Latin does not allow you to sort data on complex data types or user expressions.

Order by may not produce the same results when you rerun it. The order of tuples with the same key might change from one run to another. For example, the order of tuples with deptno 200 might change when rerun.

Look at the following code:

```
Deptorder = ORDER emp by deptno desc;
Dump deptorder;
/
(200,Radha,200000,300)
(400,Niruopam,1600000,200)
(100,Bala,100000,200)
(300,Nitya,150000,)
```

When you specify multiple fields in the Order by operator, the fields are sorted in the order specified in the code. The following code first sorts data in descending order on the deptno field and later sorts data in ascending order on ename without disturbing the order of the deptno field.

```
Deptenameorder = ORDER emp by deptno desc,ename asc;
Dump deptenameorder;

(200,Radha,200000,300)
(100,Bala,100000,200)
(400,Niruopam,1600000,200)
(300,Nitya,150000,)
```

Choosing Number of Reduce Tasks

ORDER BY launches Reduce tasks. You can choose the number of reducers using parallel.

The following code launches ten Reduce tasks using `parallel`:

```
Deptorder= ORDER emp BY deptno desc parallel 10;
```

GROUP

The `Group` operator organizes relation data into groups based on specified field names.
Here's the syntax:

```
relname= GROUP relname1 { ALL | BY fieldnames } [USING 'collected' | 'merge']
        [PARTITION BY partitioner] [PARALLEL n];
```

```
ALL            -Considers entire relation as a group.
'collected'    -Used to specify group operation only through Map task. It
                avoids reduce task.
'merge'        -Used only with cogroup to avoid reduce task.
PARTITION BY   -Used to specify Mapreduce Partitioner class
PARALLEL n     -Used to specify number of reduce tasks
```

group operations allow you to perform aggregate operations. After the group
operator, you use `foreach generate` to calculate aggregates using built-in or user-defined
functions.

The group operator produces two fields of data. One field named Group contains
data from the field upon which you performed group operations, and second is a Bag field
that contains all the columns of data from relation you have used.

The following code contains `GROUP BY` example.

```
emp = load 'employee.csv' using PigStorage(',') as (eno:int,ename:charar
ray,salary:int,dno:int);:
```

```
deptgrp = GROUP emp By dno;
describe deptgrp;
```

```
deptgrp: {group: int,emp: {(eno: int,ename: chararray,salary: int,dno: int)}}
```

ALL allows you to perform aggregate operations on an entire relation. For example,
you can get the total count of tuples in a relation and the total sum of a field in an entire
relation using it. The following code computes the total number of employees in the
relation emp:

```
Grpall = GROUP emp ALL;
Empcount = foreach grpall generate group,count(emp.eno);
Dump empcount;
(all,10000)
```

When you apply the group operator on a single field of a relation, all unique values of
that field are considered groups, and all null values are considered as one group.

When you apply the group operator on two or more fields, all unique rows on the specified fields are considered groups.

The following code groups emp relation data on the single field deptno and computes the employee count:

```
Deptgrp = GROUP emp BY deptno;
Empcount = foreach deptgrp generate group,COUNT(emp.empno);
Dump empcount;

(200,200)
(300,100)
(,11)
```

COGROUP also works the same way as the group operator. I will discuss it in Chapter 5 along with the join operator.

Using the Partitioner

The MapReduce Partitioner identifies the Reduce tasks for map output. Using the MapReduce partitioner in the Group operator helps better manage map output keys. Specify the class name along with the package name and save this class or the JAR containing this class in the lib folder.

The following code uses HashPartitioner that uses a hash function to decide the Reduce task for the Map output:

```
deptgrp = GROUP emp By deptno partition by org.apache.hadoop.mapreduce.lib.
partition.HashPartitioner;
```

Choosing Number of Reducers

The Group operator launches a Reduce task, and you can control the number of Reduce tasks to ensure optimal performance. Parallel allows you to choose a specified number of Reduce tasks.

The following code launches ten Reduce tasks:

```
Deptgrp = GROUP emp By deptno PARALLEL 10;
```

Avoiding a Reduce Task

The Reduce task is resource-intensive and can be avoided with the collected option in the group operator.

The collected option works only under a couple of conditions:

- The loader function should implement the CollectableLoadFunc interface.

- Data should be ordered on the group key.

PigStorage does not implement the CollectableLoadFunc interface. An example for such loader function is HbaseStorage.

The following code computes a column-wise employee count for the deptno column on the hbase table called employee without launching Reduce tasks:

```
emp = load 'hbase://employee' using org.apache.pig.backend.hadoop.hbase.
    HBaseStorage('empdetails:eno, empdetails:ename, empdetails:salary,
    empdetails:dno') as (empno:int,ename:chararray,salary:int,dno:int);
grpall = group emp by dno  using 'collected';
totcount =foreach grpall generate group,COUNT(test.empno);
dump totcount;
```

Stream

The Stream operator allows you to use other programs in Pig Latin code such as Hadoop streaming. You can use a shell script, Perl code, and Python code with a stream operator. The stream operator sends data as input to the program used.

Here's the syntax:

```
STREAM relationname THROUGH {`command` | programpath } [AS schema] ;
```

You will now see how to use Unix commands and a shell program with the stream operator.

Using Unix Commands

You can directly use Unix commands in the stream operator.

The following code filters tuples that contains 100 in them using the Unix grep command:

```
num = stream emp through `grep 100`;
```

Using a Shell Program

When using several Unix commands, it is better to compile all of them in a shell script and use the script in a stream operator. To make the shell script executable, use the Unix chmod command. A shell script needs to be present in the local file system.

The following code searches for 100.Cat filter100.sh:

```
Grep 100
```

The following code makes the shell script executable:

```
Chmod +x filter100.sh
```

The following code uses the shell script:

```
emp = load '/data/employee' using PigStorage(',') as (eno:int,ename:chararray,
    salary:int,deptno:int);
Emp = stream Emp through `filter100.sh`;
```

If the absolute path is not specified for shell scripts, it will look for script files in the present working directory. The stream operator opens options to use different programs in Pig Latin code. Shell script or Unix commands are easy to use and code can be easily modified as you need not compile and run like Java.

MAPREDUCE

If you have MapReduce programs already written for your project requirement, you do not need to rewrite the same functionality in Pig Latin. You can use those existing MapReduce programs in Pig Latin code using the mapreduce operator.

The following code specifies the MapReduce JAR path, and this JAR file contains the required MapReduce programs to be run. This path belongs to the local file system.

```
relname = MAPREDUCE '/path/to/mapredue.jar'
        STORE relname1 INTO 'mrinputLocation' USING storeFunc
        LOAD 'mroutputLocation' USING loadFunc AS schema
        [`params, ... `];

MAPREDUCE '/path/to/mapredue.jar'
```

This stores relation data into an input directory that is the input for the MapReduce program. This is used to integrate the input Pig Latin relation to the MapReduce program:

```
 STORE relname INTO 'mrinputLocation'
```

Once MapReduce generates output, its output will be sent back to the Pig Latin relation specified in the load operator.

```
LOAD 'mroutputLocation' USING loadFunc AS schema
```

MapReduce JAR input parameters are specified here:

```
`params`
```

Without Pig Latin, the MapReduce JAR can be run using the hadoop jar command. The following command generates the word count using hadoop-mapreduce-examples.jar. This program takes three inputs: the MapReduce program word count, the input directory names, and the output directory namesout. The following code shows how to run MapReduce code:

```
hdfs@cluster10-1:~> hadoop jar hadoop-mapreduce-examples.jar wordcount names
namesout
```

The following code runs the same MapReduce program using Pig Latin code:

```
names = load 'names';
wc =  MAPREDUCE 'hadoop-mapreduce-examples.jar' store names into
      'mrinputdir' load 'mroutputdir' as (word:chararray,num:int)
      `wordcount namesout mroutputdir` ;
dump wc;
```

The MapReduce operator launches a Hadoop job such as store and dump operators and stores output in the relation wc. MapReduce also generates data in the output directory, and you need to clean it explicitly.

CUBE

The CUBE operator allows you to perform operations such as cube and rollup.

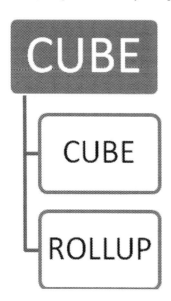

Here's the syntax:

```
relname = CUBE relname1 BY { CUBE operation| ROLLUP operation } PARALLEL n];
```

CUBE Operation   :It generates multi dimensional data set on input
                         field values.
ROLLUP Operation   : It generates multi level aggregates.
PARALLEL N   : Choosing number of reduce tasks.

CUBE

The Cube operator generates all combinations for values of the input fields given. (x,y,z) will produce a data bag as follows:

```
{ (x, y, z), (null, null, null), (x, y, null), (x, null, z),
 (x, null, null), (null, y, z), (null, null, z), (null, y, null) }
```

The cube operator is useful for generating multidimensional data sets in Pig Latin.

The schema of a cube is similar to a schema of a group. The cube schema will have a group tuple and a cube bag.

```
enoename = cube emp by CUBE(eno,ename);
describe enoename;
enoename: {group: (eno: int,ename: chararray),cube: {(eno: int,ename:
chararray,salary: int,dno: int)}}
```

The CUBE operator launches a Reduce task so you can choose the number of Reduce tasks using a parallel keyword. The following code launches ten Reduce tasks to perform a cube operation:

```
enoename = cube emp by CUBE(eno,ename) parallel 10;
```

ROLLUP

The ROLLUP operator produces a data bag with a hierarchy of input values. (x,y,z) will produce a data bag like the following:

```
{ (x, y, z), (x, y, null), (x, null, null), (null, null, null) }
```

The following code generates a rollup on the employee number and employee name fields:

```
Rollupex = CUBE emp by ROLLUP(eno,ename);
```

As it launches the Reduce task, you can choose the number of Reduce tasks to be executed, as shown here:

```
Rollupex = CUBE emp by ROLLUP(eno,ename) parallel 10;
```

With the Cube operator, you complete learning most of Pig Latin operators. Now you will see how to write dynamic scripts using the parameter substitution feature of Pig Latin.

Parameter Substitution

Parameter substitution allows you to declare a variable or parameter within the Pig Latin script whose value you can provide at the time of running the script. Parameter substitution allows you to write more dynamic scripts. This feature is useful in a production environment where you do not change an already working script very frequently.

The following code returns five tuples from the employee relation, and the employee relation will load data from the input directory /data/employee:

```
emp = load '/data/employee' using PigStorage(',')  as (eno:int,ename:chararray,
     salary:int,deptno:int) ;
emp5 = limit emp 5;
dump emp5;
```

If you want to have five employee tuples from some other input directory, you need to modify the load statement and run the Pig script. You need to avoid frequent modifications as changes in requirements will be a common thing and you need to write dynamic script to make reusable for as many changes as possible. Pig Latin provides a feature called *parameter substitution* that will help you write dynamic and reusable Pig Latin scripts.

Parameter substitution can be achieved in two ways, as listed in Figure 4-2.

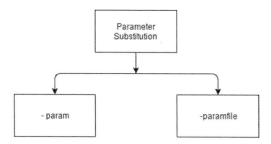

Figure 4-2. *Parameter substitution*

Now you will learn how these two options help you write dynamic scripts.

-param

-param allows you to provide values at run time for declared variables in a Pig Latin script. This is two-step process.

1. First you need to define a parameter using a dollar symbol ($) within the Pig Latin code.

The following code avoids hard-coding the input directory by defining a parameter called `inputdir`:

```
emp = load '$inputdir' using PigStorage(',')  as (eno:int,ename:chararray,
    salary:int,deptno:int);
```

Without the parameter substitution feature, you would hard-code the input path as follows, which needs to be changed every time the input path is changed.

```
emp = load '/data/employee' using PigStorage(',')  as (eno:int,ename:chararr
ay,salary:int,deptno:int);
```

> 2. The value of parameter `inputdir` will be substituted at the time of running the script.

The following code substitutes the `inputdir` parameter with `/data/employee`. The parameter name and its value need to be specified after the `-param` option.

```
Pig -param inputdir=/data/employee -f dumpemp5.pig
```

Now you can run the Pig Latin script with the input directory of your choice without altering the code.

-paramfile

The problem with the `-param` option is that the number of `-param` options depend on the number of variables defined in a script that makes it lengthy, as follows:

```
Pig -param inputdir=/data/employee -param indate=10-01-2016 –delim=:
-outdir=/data/outemp -f dumpemp5.pig
```

The previous command lacks readability and is also hard to maintain.

To address this problem, you can use the `-paramfile` option. After defining parameters in a Pig Latin script, you need to create a file that contains parameter names and their values.

The following parameter file contains two parameters called `inputdir` and `outputdir`, as well as their values:

```
Cat /home/hdfs/inputoutput.properties
Inputdir=/data/employee
Outputdir=/data/out/emp
```

You will specify the parameter file path when running the script.

The following example makes a command readable and also makes it much easier to maintain variables using a separate file:

```
Pig -parmafile /home/hdfs/inputoutput.properties -f dumpemp5.pig
```

Summary

In this chapter, you learned three important fundamentals of Pig Latin.

- How to run Pig Latin code using `pig -e`, `pig -f`, the Grunt shell, and Java code.

- The purpose and usage of Pig Latin basic operators such as `limit`, `union`, `split`, `load`, `store`, and so on.

- How to make Pig Latin script more dynamic using the `-param` and `-paramfile` options.

CHAPTER 5

Joins and Functions

Many times you need to retrieve data from more than one relation to generate more meaningful and readable reports. You can use the joins feature of Apache Pig to retrieve data from more than one relation.

In Pig Latin, joins can be two types: equi joins and non-equi joins. Equi joins retrieve data from more than one relation applying equal conditions. A non-equi join retrieves data from more than one relation applying conditions that are not equal.

Equi joins can be further categorized into inner joins and outer joins. Inner joins return only matching rows, and outer joins return both matching and nonmatching rows. Nonmatching rows can be from left, right, or both. Outer joins can be further categorized into left outer join, right outer join, and full outer join.

Figure 5-1 shows all the join types.

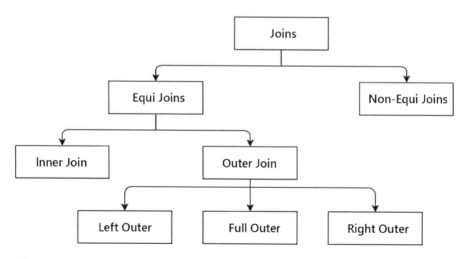

Figure 5-1. *Join types*

© Balaswamy Vaddeman 2016
B. Vaddeman, *Beginning Apache Pig*, DOI 10.1007/978-1-4842-2337-6_5

Join Operators

You will now learn how to perform join operations using Pig Latin. Pig Latin provides two operators, join and cogroup, to perform join operations.

Consider the following employee and department data:

Employee				
Emp No	Emp Name	Designation	Salary	DeptNo
10001	Bala	Senior Software Engineer	1000000	200
1002	Radha	HR Executive	450000	300
1003	Nirupam	Tech Lead	1500000	200
1004	Nitya	Software engineer	350000	

Department	
DeptNo	Department Name
200	Customer Services
300	Human Resource
400	Administration

Equi Joins

Equi joins use an equality condition to retrieve data from more than one table. To retrieve the department name of an employee, you would have to match the department number of the employee relation with the department number of the department relation.

Inner Joins

An inner join retrieves only those rows that are matched. For example, some employees may not have a department number. Such employee rows are not displayed in the output. To perform an inner join, you use relations with a common column.

The following script retrieves the matching rows in the Employee and Department relations using the common column deptNo:

```
Employee = load '/user/hdfs/in/employee' using PigStorage(',') as
          (eno:int,ename:charaaray,desg:chararray,deptno:int);
Department=load '/user/hdfs/in/dept' using PigStorage(',') as
          (deptno:int,dname:chararray);
joinempdept = join Employee by deptno,Department by deptno;
enamedanme = foreach joinempdept generate ename,dname;
dump enamedname;
```

The following output does not display the Nitya row in the `Employee` relation because it does not have a `deptno` value. The administration row from the `Department` relation is also not displayed because it does not have a matching row.

Bala	Customer Services
Radha	Human Resource
Nirupam	Customer Services

Outer Joins

As mentioned earlier, outer joins retrieve not only matching rows but also nonmatching rows from relations. Outer joins can be performed in three ways.

Left Outer Join

Left outer joins retrieve matching rows including nonmatching rows from the left relation. The first relation used in a script is considered the left relation, and the second one is considered the right relation. You can use only `LEFT` or `LEFT OUTER` in a script to specify it as a left outer join.

The following code also retrieves the nonmatching row from the left relation, which is the Nitya row:

```
joinempdept = join Employee by deptno LEFT OUTER,Department by deptno;
```

Bala	Customer Services
Radha	Human Resource
Nirupam	Customer Services
Nitya	

Right Outer Join

The right outer joins retrieve matching rows including nonmatching rows from the right relation. You can use only `RIGHT` or `RIGHT OUTER` in the script to specify it as a right outer join.

The following code retrieves the nonmatching row from the right-side relation, which is the administration row:

```
joinempdept = join Employee by deptno RIGHT OUTER,Department by deptno;
```

Bala	Customer Services
Radha	Human Resource
Nirupam	Customer Services
	Administration

Full Outer Join

You can retrieve nonmatching rows from both relations using a full outer join. You can use only FULL or FULL OUTER in a script to specify it as a full outer join.

The following code retrieves nonmatching rows from both relations:

```
joinempdept = join Employee by deptno FULL OUTER,Department by deptno;
```

Bala	Customer Services
Radha	Human Resource
Nirupam	Customer Services
Nitya	
	Administration

All join operations launch Reduce tasks, and Reduce tasks are costly operations in Hadoop in terms of runtime. Hence, you must optimize joins for better performance. I will discuss join optimizations in Chapter 16.

cogroup

cogroup is primarily used for achieving join functionality, but it works in much the same way as the GROUP operator. cogroup groups data from multiple relations based on a common column. The difference between the group and cogroup operators is that the GROUP operator groups data based on columns from one relation, and cogroup groups data on common columns from two or more relations.

The difference between the cogroup and join operators is that cogroup results will include a common column, the data from the left relation, and the data from the right relation, whereas join results will include data only from the left relation and the right relation.

The following code performs an outer join on the Employee and Department relations using the common column deptno:

```
emp = load '/data/employee' using PigStorage(',') as
      (empno:int,ename:chararray,salary:int,deptno:int);
dept = load 'dept.csv' using PigStorage(',') as (deptno:int,dname:chararray);
cogrp = cogroup emp by deptno outer,dept by deptno;
describe cogrp;
```

```
{group: int,emp: {(empno: int,ename: chararray,salary: int,deptno: int)},
dept: {(deptno: int,dname: chararray)}}
```

The describe operator output after the join will look like this:

```
Empdeptjoin = join emp by deptno,dept by deptno;
Describe empdeptjoin;
empdeptjoin: {emp::empno: int,emp::ename: chararray,emp::salary: int,
              emp::deptno: int,dept::deptno: int,dept::dname: chararray}
```

CROSS

The cross operator generates the Cartesian product of two or more relations. This operator can be used for non-equi joins in Pig Latin.

At times you may have to retrieve data from multiple relations that may not have any common column. For example, assume that in addition to emp and dept, you have another relation called salaryrange that defines minsalary, maxsalary, and rangename for that salary range.

Salary range		
Min salary	Max salary	rangename
0	500000	LOW
500001	1000000	MEDIUM
1000001	5000000	HIGH
5000000	10000000	VERY HIGH

To classify employee salary into a range, you must retrieve data from the employee and salaryrange relations, and these two do not have any common column. One solution is to generate a Cartesian product of both relations that matches one row from one relation with every row of the other relation. The Cartesian product of the employee and salaryrange relations generates 16 (4*4) rows. Apply the between operation on the output to get the salary range for an employee salary. This is one example of a non-equi join. You can use the CROSS operator to generate the Cartesian product of two or more relations.

The following code generates the Cartesian product for the employee and salaryrange relations:

```
Cartesianprod= cross employee,salaryrange;
```

Complete the previous code to generate the employee name and range name, as shown here:

ENAME	RANGE NAME
Bala	MEDIUM
Radha	LOW
Nirupam	HIGH
Nitya	LOW

Functions

Pig Latin provides several functions to process data. The two types of functions in Pig Latin are built-in functions and user-friendly functions. Built-in functions come with Pig, and you can directly use them by mentioning their name. If you cannot find a suitable function, you can write your own function and use it in Pig Latin. Such functions are called *user-defined functions*.

Pig Latin provides the built-in functions listed in Figure 5-2. Built-in functions are case sensitive, and most of them are in uppercase.

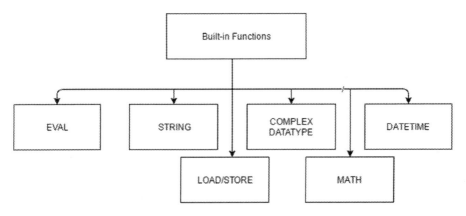

Figure 5-2. Built-in functions

String Functions

String functions are used to work on a sequence of characters. Most string functions work the way string functions work in Java. The following are some String functions.

UPPER

The UPPER function converts input string into uppercase. It allows chararray data.

The following code converts all employee names into uppercase:

```
ucase= foreach employee generate UPPER(ename);
```

LOWER

The LOWER function converts input string into lowercase. The LCFIRST function converts the first letter into lowercase, and UCFIRST converts the first letter into uppercase.

The following code converts the first letter of all employee names into uppercase:

```
firstucase= foreach employee generate UCFIRST (ename);
```

TRIM

Extra spaces are a common issue in data. The field data might start with a space, might end with space, or both. TRIM removes both spaces. If you are sure that data only starts with space or only ends with space, then you can use LTRIM or RTRIM, respectively. The TRIM operation takes more time than LTRIM and RTRIM. Sometimes joins do not work because of space issues, so it is a good idea perform trim operations on common columns.

The following code removes spaces from both ends of the string TEST:

```
Nospace = foreach dummy generate TRIM (' TEST ');
```

REPLACE

The REPLACE function is used to replace existing characters with new specified characters. It internally uses Java's replaceAll function, as follows: http://docs.oracle.com/javase/6/docs/api/java/lang/String.html#replaceAll(java.lang.String, java.lang.String). Java escape characters replace special characters. It also allows RegEx like Java functions.

Here's the syntax:

```
REPLACE(string, 'existingStr', 'newChars');
```

The following code replaces the asterisk characters with the word *star*:

```
foreach dummy generate REPLACE('character *','\\*','star');
```

STRSPLIT

The STRSPLIT function splits the string using the delimiter specified. It works like a string tokenizer. A delimiter can also be a regular expression. limit is the integer value that specifies the number of times the delimiter must be applied.

Here's the syntax:

```
STRSPLIT(string, delim, limit)

Tokens= foreach names generate STRSPLIT('this:is:test:',':',2);
```

The STRSPLITTOBAG function is similar to STRSPLIT. The only difference is that STRSPLITTOBAG returns a bag data type. For example, the previous code with STRSPLIT returns the ((this,is:test:)) output and with STRSPLITTONAG returns the ({(this),(is:test:)}) output.

UniqueID

The UniqueID function returns a unique value to every record. It uses taskindex and sequence to generate a unique value.

SUBSTRING

The SUBSTRING function returns a substring of a string. It takes three arguments. They are the input string, start index, and stop index. SUBSTRING returns characters starting from the start index until the stop index.

The following code returns the World substring from "Hello World":

```
Foreach dummy generates SUBSTRING('Hello World','6','11')
```

Mathematical Functions

Most mathematical functions in Pig Latin work the same way as Java mathematical functions.

Let's look at some mathematical functions in Pig Latin.

FLOOR

The Floor function returns the largest previous integer that is less than or equal to the input number like a mathematics floor function.

The following are some examples:

Floor with input	Output
floor(4.9)	4
floor(4)	4
floor(-4.9)	-5

```
Floorvalue=Foreach dummy generate floor(4.9)
```

CEIL

The CEIL function returns the smallest next integer that is greater than or equal to the input number like a mathematics ceil function.

Here's an example:

CEIL with input	Output
CEIL(4.1)	5
CEIL(4)	4
CEIL(-4.1)	-4

```
Ceilvalue = Foreach dummy generate ceil(4.9)
```

ROUND

The ROUND function returns a rounded value of an input number like a mathematical round function. It is performed on decimals.

Here's an example:

Round with inpu number	Output
ROUND(4.1)	4
ROUND(4.5)	5
ROUND(4,0)	4
ROUND(-4.1)	-4
ROUND(-4.5)	-5

If you have more than one decimal value and want to preserve a specific number of digits, you can use the ROUND_TO function. The ROUND_TO function allows you to specify the number of digits to be retained as a second argument.

```
digits4=Foreach dummy generate Round_To(123.22,4)
```

RANDOM

The RANDOM function returns a pseudo-random number between 0.0 and 1.0.

```
Randnum = foreach dummy generate RANDOM();
```

ABS

The ABS function returns the absolute value of an input number.

```
Abvsvalue = forach dummy generate ABS(-4.3);
```

Pig Latin provides more mathematical functions such as sin, tan, and cos. Please refer to Appendix A for a complete list of functions.

Date Functions

Pig Date functions depend on the Java Date API and the JODATIME API.
The following are some important functions in Pig Latin.

CurrentTime

The CurrentTime function returns the current datetime object similar to the JODATIME API. The datetime object provides the date, time, and time zone information.
The following code returns the current date with the time and an India time zone:

```
Timenow = forach dummy generate CurrentTime();
Dump timenow;
(2016-06-27T13:01:39.402+05:30)
```

GetDay

The GetDay function returns the day from the input datetime field.
The following code returns the current day from the current date:

```
Currentday=foreach names generate GetDay(CurrentTime());
```

Pig provides similar functions to get the hour, second, week, month, and year using the GetHour, GetSecond, GetWeek, GetMonth, and GetYear functions.

DAYSBETWEEN

DAYSBETWEEN returns the number of days between two datetime fields.

The following code returns 9 as the day difference between 2016-06-19 and 2016-06-10:

```
days = foreach dummy generate DaysBetween(ToDate('2016-06-19'),
    ToDate('2016-06-10'));
```

Pig also provides similar functions such as HOURSBETWEEN, MINUTESBETWEEN, SECONDSBETWEEN, WEEKSBETWEEN, MONTHSBETWEEN, and YEARBETWEEN to find the difference in hours, minutes, seconds, weeks, months, and years.

TODATE

TODATE converts the input field value to a date format. This function can be used in four ways.

The following code converts milliseconds to the date format; milliseconds are considered from 1970-01-01T00:00:00.000Z. You can use the ToMilliSeconds function to convert the date to milliseconds.

```
ToDate(milliseconds)
```

The following code returns the current date in milliseconds:

```
currentmillis = foreach dummy  generate ToMilliSeconds(CurrentTime());
```

```
ToDate(isostring)
```

It converts ISOString to a datetime format.

The following code converts the string data into the datetime data type:

```
strtodate = foreach dummy generate ToDate('2016-06-19');
```

```
ToDate(string, requiredformat)
```

The following code converts the string to a user-specified format. You can use the Java SimpleDateFormat class specification (http://docs.oracle.com/javase/6/docs/api/java/text/SimpleDateFormat.html) to set the format.

```
ToDate(string, requiredformat, timezone)
```

It converts a string to the user-specified format and time zone.

You can use the ToString function to convert the datetime field value to the string type.

TOUNIXTIME

The ToUnixTime function returns the Unix time for the datetime field value. It is the number of seconds elapsed since January 1, 1970, 00:00:00.000 GMT.

The following code returns the current date in seconds:

```
currentseconds = foreach dummy generate ToUnixTime(CurrentTime());
```

I will discuss user-defined functions (UDFs) in greater detail in Chapter 11.

EVAL Functions

All functions that extend the EVALFUNC class are EVAL functions, and they run once per tuple.

The following are some EVAL functions.

AVG

The AVG function returns the average of numbers from the input field. The AVG function must be preceded by the GROUP operator.

The following code returns the department-wise average salary:

```
deptgrp = GROUP emp by deptno;
flat = foreach deptgrp generate FLATTEN(group),AVG(emp.salary);
```

MIN

The MIN function returns smallest number from the input field. The MIN function needs to be preceded by the GROUP operator. It ignores null values. It returns the department-wise minimum salary.

```
deptgrp = GROUP emp by deptno;
flat = foreach deptgrp generate FLATTEN(group),MIN(emp.salary);
```

Like with MIN, Pig Latin also provides the MAX function that returns the largest number from the input field.

COUNT

The COUNT function returns the total number of elements from the input field. The COUNT function must be preceded by the GROUP operator. The COUNT function ignores null values. To consider null values, use COUNT_STAR.

The following code returns the employee count of every department number:

```
deptgrp = GROUP emp by deptno;
flat = foreach deptgrp generate FLATTEN(group),COUNT_STAR(emp.deptno);
```

BagToString

The BagToString function converts data from a bag structure to a normal string.

Complex Data Type Functions

Here are the complex data type functions.

TOTUPLE

The TOTUPLE function converts given fields into the tuple data type.

The following code converts four fields (empno, ename, salary, and deptno) into a tuple:

```
grunt>emptuple= foreach emp generate TOTUPLE(empno,ename,salary,deptno);
grunt>dump emptuple;

((100,Bala,100000,200))
((200,Radha,200000,300))
((300,Nitya,150000,))
((400,Nirupam,1600000,200))
```

TOBAG

The TOBAG function converts given fields into the bag data type.

The following code converts four fields (empno, ename, salary, and deptno) into a bag data type:

```
empbag= foreach emp generate TOBAG(empno,ename,salary,deptno);
dump empbag;

({(100),(Bala),(100000),(200)})
({(200),(Radha),(200000),(300)})
({(300),(Nitya),(150000),()})
({(400),(Nirupam),(1600000),(200)})
```

TOMAP

The TOMAP function converts given fields into the map data type. As map contains key and value data, you provide two fields as input and the key as a chararray data type.

The following code converts ename and empno fields into a map data type:

```
Grunt>empmap= foreach emp generate TOMAP(ename,empno);
grunt>dump empmap;

([Bala#100])
([Radha#200])
([Nitya#300])
([Nirupam#400])
```

TOP

The TOP function returns the top *N* tuples from a bag.

```
Top(N,tuplenumber,relationname)
```

The TOP function takes three inputs: the number of tuples to be returned, the tuple number at which counting begins, and the relation name from which tuples must be retrieved.

The following code generates the deptno-wise employee count and then returns the first tuple from the output. You have created an alias for the TOP function and grouped deptno-wise employee count data. You must apply the GROUP operator before applying the TOP function.

```
DEFINE desca TOP('DESC');---first define macro

emp = load 'employee.csv' using PigStorage(',') as
      (empno:int,ename:chararray,salary:int,deptno:int);
deptnogrp = group emp by deptno;
empcount = foreach deptnogrp generate FLATTEN(group),COUNT(emp) as ecount;
empcount = GROUP empcount BY ecount; ---group before applying top
empcount = foreach empcount{
result = desca(2,1,empcount);
 generate result;
}
dump empcount;
```

Load/Store Functions

Pig Latin provides several functions for reading and writing different data formats such as ORC, Binary, and Avro, and some functions are used for reading and writing different technology data such as HBase and Accumulo.

You will learn some load/store functions here.

JsonLoader/JsonStorage

JsonStorage writes given relation data into the file system in JavaScript Object Notation (JSON) format. It also generates a hidden schema file called .pig_schema containing field names and data types and a hidden header file called .pig_header in the output directory. The header file contains the field names.

The following code writes employee relation data in JSON format:

```
employee = load 'employee.csv' using PigStorage(',') as
           (eno:int,ename:chararray,salary:int,deptno:int);
store employee into 'csvtojson'using JsonStorage();
```

It generates the following files in the output directory:

```
ls -ltra csvtojson
total 36
-rw-r--r-- 1 hdfs hadoop     8 Jun 23 02:44 ._SUCCESS.crc
-rw-r--r-- 1 hdfs hadoop     0 Jun 23 02:44 _SUCCESS
-rw-r--r-- 1 hdfs hadoop    12 Jun 23 02:44 ..pig_schema.crc
-rw-r--r-- 1 hdfs hadoop   429 Jun 23 02:44 .pig_schema
-rw-r--r-- 1 hdfs hadoop    12 Jun 23 02:44 ..pig_header.crc
-rw-r--r-- 1 hdfs hadoop    24 Jun 23 02:44 .pig_header
-rw-r--r-- 1 hdfs hadoop    12 Jun 23 02:44 .part-m-00000.crc
-rw-r--r-- 1 hdfs hadoop   232 Jun 23 02:44 part-m-00000
drwxr-xr-x 2 hdfs hadoop  4096 Jun 23 02:44 .
drwxr-xr-x 7 hdfs hadoop  4096 Jun 23 02:57 ..
```

JsonLoader reads JSON data. JsonLoader can be used with or without a schema. If you do not specify the schema in code, it assumes there is a schema file called .pig_schema in the input directory.

```
employee = load 'employee.csv' using JsonStorage(',')
dump employee;
```

```
{"eno":100,"ename":"Bala","salary":100000,"deptno":200}
{"eno":200,"ename":"Radha","salary":200000,"deptno":300}
{"eno":300,"ename":"Nitya","salary":150000,"deptno":null}
{"eno":400,"ename":"Niruopam","salary":1600000,"deptno":200}
```

PigStorage

The PigStorage function reads and writes structured text files. It is the default load/store function in Pig Latin. It takes two inputs; one is the delimiter, and the other is a list of options. The default delimiter is a tab (\t). The delimiter is a single character, and if control characters such as Ctrl+B are delimiters, you need to use their Unicode character (\u002) as the delimiter.

```
PigStorage('[delimiter]','[options]');
```

The ('schema') option is used for reading data using a schema from the schema files from their Unicode character (\u002) as the input directory and for writing the schema to the output directory.

The following code generates a schema file while writing data to the output directory:

```
store emp into 'empnew' using PigStorage(',','-schema') ;
```

You can use the ls -ltra command to view hidden schema files and header files.

```
hdfs@cluster10n1:~> ls -ltr emp
-rw-r--r--  1 hdfs hadoop      8 Jun 27 13:15 ._SUCCESS.crc
-rw-r--r--  1 hdfs hadoop      0 Jun 27 13:15 _SUCCESS
-rw-r--r--  1 hdfs hadoop     12 Jun 27 13:15 ..pig_schema.crc
-rw-r--r--  1 hdfs hadoop    147 Jun 27 13:15 .pig_schema
-rw-r--r--  1 hdfs hadoop     12 Jun 27 13:15 ..pig_header.crc
-rw-r--r--  1 hdfs hadoop      3 Jun 27 13:15 .pig_header
-rw-r--r--  1 hdfs hadoop     12 Jun 27 13:15 .part-m-00000.crc
-rw-r--r--  1 hdfs hadoop    120 Jun 27 13:15 part-m-00000
drwxr-xr-x  2 hdfs hadoop   4096 Jun 27 13:15 .
drwxr-xr-x 11 hdfs hadoop  12288 Jun 27 13:17 ..
```

The following are the options:

- ('noschema'): This is used to ignore the available schema file.

- ('tagsource'): tagsource is outdated; tagpath is the latest option. This adds the first column with the input path.

- ('tagPath'): This adds the pseudo-column INPUT_FILE_PATH to the beginning of the record.

- ('tagFile'): This adds the pseudo-column INPUT_FILE_NAME to the beginning of the record.

TextLoader

The TextLoader function is used for reading unstructured data. It is used only for reading data from the source and does not work for writing data.

HbaseStorage

The HbaseStorage function is used for reading and writing HBase table data. The HbaseStorage function takes two inputs; one is a list of columns, and the other consists of the options to be used. The load operator takes the table name in hbase://tablename format.

Here's the syntax:

```
Relname = load 'hbase://tablename' using HbaseStorage('columnnames',['options']);
```

Specify the column name and column family with a colon (:) between them. To delimit columns, use either a space or a comma.

Some important features of HbaseStorage are discussed here:

- You can specify all columns in a column family using * or without specifying any column name. This will produce a Pig map with column names as a key. The following code retrieves all columns from the empdetails column family from the employee HBase table.

  ```
  emp = load 'hbase://employee' using org.apache.pig.backend.
      hadoop.hbase.HBaseStorage('empdetails:*') ;
  dump emp;

  ([ename#bala,dept#200,salary#100000,eno#100])
  ```

- You can specify column names that start with a particular string. The following code retrieves only columns that start with e. This will also produce data in a Pig map data type.

  ```
  emp = load 'hbase://employee' using org.apache.pig.backend.
      hadoop.hbase.HBaseStorage('empdetails:e*') ;
  dump emp;
  ([ename#bala,eno#100])
  ```

- You can specify a list of column names without any filter condition using a space or a comma as a delimiter. The following code retrieves eno, ename, salary, and deptno columns from the employee table. You can define any schema using this method.

  ```
  emp = load 'hbase://employee' using org.apache.pig.backend.
      hadoop.hbase.HBaseStorage('empdetails:eno,empdetails:ename,
      empdetails:salary,empdetails:deptno') as (eno:int,ename:chararray,
      salary:int,deptno:int);
  dump emp;
  ```

The following are some options of HbaseStorage that you can use:

- -loadKey retrieves the row key as the first value in every tuple, and the default value is false.

- -gt=minrowKeyVal retrieves rows with a rowKey greater than minrowKeyVal.

- -lt=maxrowKeyVal retrieves rows with a rowKey less than maxrowKeyVal.

- • -regex=regex retrieves rows that match this RegEx on a row key.

- • -limit=numRowsPerRegion retrieves the maximum numRowsPerRegion number of rows per region.

- • The -delim=delim delimiter can be used on a column list (the default is whitespace).

- • -minTimestamp=timestamp returns cell values that have a creation timestamp greater than or equal to the timestamp value specified.

- • -timestamp=timestamp retrieves cell values that have a creation timestamp equal to the specified value.

- • -includeTimestamp displays the timestamp after the rowkey (for example, rowkey, timestamp, …).

Refer to Appendix C for a complete list of options.

The following code retrieves the row key along with the columns eno, ename, salary, and deptno:

```
emp = load 'hbase://employee' using org.apache.pig.backend.hadoop.hbase.
    HBaseStorage('empdetails:*','-loadKey=true')
dump emp;
```

Storing Data into HBase

To store Pig relation data in the HBase table, specify the HBase table in a clause and a list of columns in the HbaseStorage() function. The first column is considered as a row key. The table must be present in HBase.

The following code stores data in the Hbase table emp and the first column eno as a row key:

```
emp = load 'employee.csv' using PigStorage(',') as
    (empno:int,ename:chararray,salary:int,deptno:int);
store emp into 'hbase://emp' using org.apache.pig.backend.hadoop.hbase.
HBaseStorage(
    'empdetails:ename empdetails:salary empdetails:deptno ');
```

If you scan the emp table in HBase, you will see eno stored as a row key.

```
hbase(main):001:0> scan 'emp'
ROW                                          COLUMN+CELL
 100                                         column=empdetails:deptno,
timestamp=1466921682193, value=200
 100                                         column=empdetails:ename,
timestamp=1466921682193, value=Bala
 100                                         column=empdetails:salary,
timestamp=1466921682193, value=100000
1 row(s) in 0.3040 seconds
```

Like HbaseStorage, Pig Latin also provides ACCUMULOSTORAGE functions to read and write data from and to the Accumulo data store.

OrcStorage

The OrcStorage function is used for both reading and writing (Optimized Row Columnar (ORC) data format. ORC is an efficient data format used for high performance.

Here's the syntax:

```
Relname = load '/path/to/dataset' using OrcStorage(['options']);
```

You can specify compression and stripe size as options. Refer to Appendix C for a complete list of options.

Options are available while writing data, and the following code converts CSV data into ORC format applying Snappy compression:

```
employee = load 'employee.csv' using PigStorage(',') as
           (eno:int,ename:chararray,salary:int,deptno:int);
store employee into 'csvtoorc'using OrcStorage('-c SNAPPY');
```

Loading Data

If filter conditions are available after the load operator and they are on normal data types, they will be moved to the load operator for better performance.

The following code loads ORC data using the OrcStorage function, and the filter condition deptno=200 is moved to the load operator so data will be loaded more quickly.

```
employee = load 'csvtoorc' using OrcStorage as
           (eno:int,ename:chararray,salary:int,deptno:int);
depl200 = filter emp by deptno==200;
```

Summary

In this chapter, you learned two important features of Pig Latin.

- How to perform join operations such as equi joins and nonequi joins using the join, cogroup, and cross operators

- How to use different built-in functions of Pig Latin like String, Math, Date, Eval, and Load/Store

CHAPTER 6

░ ░ ░

Creating and Scheduling Workflows Using Apache Oozie

Big data processing in Hadoop usually involves multiple technologies that have to be implemented in a certain order and manner. Often, these technologies also interact with one another. For instance, a certain step n in the workflow can be executed if and only if step n-1 has been successfully executed. Manually executing each of these multiple steps is time-consuming. Apache Oozie addresses this problem by providing dependency management among different steps and technologies.

Apache Oozie is a web application that provides a job execution service for Hadoop ecosystem jobs. It can execute both order-based jobs and time-based jobs. Currently, it supports MapReduce, Hive, Sqoop, Spark, and Pig jobs. You can also run cascading jobs. In this chapter, you will learn how to submit Pig jobs using Oozie. (*Oozie* means "elephant rider" or "elephant keeper" in Burma.)

Types of Oozie Jobs

Primarily, Oozie can execute two types of jobs: workflow jobs and coordinator jobs. Workflow contains a bunch of Hadoop ecosystem jobs that run as a single unit in a specified order. Coordinator jobs provide a scheduler service for Hadoop ecosystem jobs. They can schedule Hadoop jobs depending on both a time and an event. You can launch a list of coordinator jobs as a bundle application in Oozie. In this chapter, you will learn how to write simple workflow and coordinator applications.

Workflow

To write a workflow application, you will need a `job.properties` file, a `workflow.xml` file, and a Hadoop ecosystem job script file. For example, to run a Pig job after a Hive job, you will require a `job.properties` file, a `workflow.xml` file, a Pig script file, and a Hive script file. Now you will learn more about these files.

© Balaswamy Vaddeman 2016
B. Vaddeman, *Beginning Apache Pig*, DOI 10.1007/978-1-4842-2337-6_6

job.properties

The job.properties file in a workflow application primarily contains Hadoop environment details and workflow details such as the HDFS namespace, the job tracker URL, the workflow application path, and some other details.

Use the fs.defaultFS property to specify the HDFS namespace. This property is available in core-site.xml, and the value can be taken from that file.

The following is an example that specifies the HDFS namespace:

```
fs.defaultFS=hdfs://cluster10
```

In older versions, you specified the namenode address of the Hadoop cluster as follows:

```
nameNode=hdfs://namenodehostname:8020
```

Use the property mapreduce.jobtracker.address to specify the jobtracker address of the Hadoop cluster. This property is available in the mapred-site.xml file.

```
mapreduce.jobtracker.address=cluster10-1:8021
```

Use the property oozie.wf.application.path to specify the workflow application path in the HDFS cluster and use the oozie.libpath property to specify the library path.

The job.properties file is stored in the local file system and can be easily modified to run the Oozie script on different clusters.

workflow.xml

The workflow is defined using workflow.xml. It contains two types of XML tags also called *nodes*: control nodes and action nodes.

Control nodes specify the job priority, that is, which job has to be run first and which job has to be run last. Some control nodes include the following:

- *Start node*: Decides which Hadoop job has to be run first

- *End node*: Decides which Hadoop job has to be run last

- *Kill node*: Aborts a Hadoop job when it fails

The action node decides the job to be executed. It supports MapReduce, Hive, Pig, Sqoop, Ssh, Distcp, and Spark jobs. Every job either fails or succeeds; hence, every action will have two more tags called OK and Error to decide the next action. The OK tag decides the next job to be run if the current job is successful, and the Error tag will decide the next action if the current job fails.

Figure 6-1 depicts how different nodes work. The action is executed until it fails or it reaches the end node.

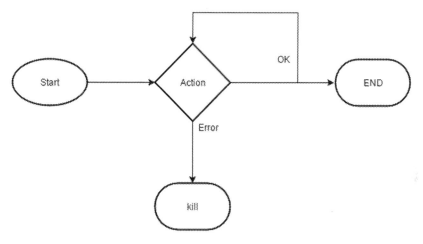

Figure 6-1. How nodes work

Using a Pig Latin Script as Part of a Workflow

The following are the steps for using a Pig Latin script as part of a workflow.

Writing job.properties

First, you must write the job.properties file. As discussed, the jobtracker address, the file system address, and the workflow application path must be set in the job.properties file. You can declare variables with some value so that they are reusable within the Oozie application.

The following sample job.properties file defines variables called examplesRoot with value examples. It also has a variable called user.name that gets resolved to the current user of the operating system. If the current user is hdfs, then oozie. wf.application.path will resolve to /user/hdfs/examples/apps/pig. You must create this directory in HDFS.

```
fs.defaultFS=hdfs://cluster10
mapreduce.jobtracker.address=cluster10n1:8021
queueName=default
examplesRoot=examples
oozie.use.system.libpath=true
oozie.libpath=/user/${user.name}/${examplesRoot}/apps/examples-lib
oozie.wf.application.path=/user/${user.name}/${examplesRoot}/apps/pig
```

workflow.xml

workflow.xml starts with the <workflow-app> XML tag, and you can use the attribute name in it to name your Oozie workflow. The <start> tag specifies the action name to be run first in the workflow using the attribute called to.

The workflow.xml file shown next contains an action named first-pig-node to be started first. The <start> tag is followed by a list of <action> tags representing action nodes. Every <action> tag has <ok> and <error> tags to represent success and failure. The tags <ok> and <error> have two attributes to represent the next tags to be executed. The tag <end> represents the end of the workflow. The <kill> tag represents error message to be thrown in case an action fails.

The Pig action should have <pig> tag within it. The <script> tag contains the path of the Pig script file that must be run. This script file must be placed in the hdfs directory mentioned in the job.properties file as the value of oozie.wf.application.path. The parameters ${jobTracker} and ${nameNode} will resolve using their values defined in the job.properties file.

A sample workflow.xml file follows:

```
<workflow-app xmlns="uri:oozie:workflow:0.2" name="pig-wf">
    <start to="first-pig-node"/>
    <action name="first-pig-node">
        <pig>
            <job-tracker>${jobTracker}</job-tracker>
            <name-node>${nameNode}</name-node>
<script>id.pig</script>
        </pig>
        <ok to="end"/>
        <error to="fail"/>
    </action>
    <kill name="fail">
        <message>Pig failed, error message[${wf:errorMessage(wf:lastErrorNo
            de())}]</message>
    </kill>
    <end name="end"/>
</workflow-app>
```

Set a Value to a Property

You can use the <configuration> tag to specify a value to a property. The <configuration> tag has a property tag, and it in turn has two tags, <name> and <value>.

The following code sets the MapReduce queue name to the default:

```
<configuration>
                <property>
                    <name>mapred.job.queue.name</name>
                    <value>default</value>
                </property>
</configuration>
```

Passing Parameter Values

You can pass values to parameters such as the -param option in Pig. You can use the <param> tag immediately after the <script> tag to pass a value to a parameter. The following code sets the value /user/data/input to the parameter INPATH:

```
<param>INPATH=/user/data/input </param>
```

You can write the initialization steps using a <prepare> tag that is to be executed before the workflow is launched.

The following code deletes a folder before the workflow begins:

```
<prepare>
                <delete path="/user/hdfs/output-data/pig"/>
</prepare>
```

Uploading Files to HDFS

After you write job.properties and workflow.xml, you can upload workflow.xml and the Pig script file to the HDFS directory specified as a value of oozie.wf.application. path. You also need to upload JARs to the HDFS directory specified in the lib.path property of the job.properties file.

```
Hdfs dfs -put /path/to/workflow.xml /user/hdfs/examples/apps/pig
Hdfs dfs -put /path/to/script.pig /user/hdfs/examples/apps/pig
```

Submit the Oozie Workflow

You can use the Oozie command-line interface to submit the Oozie workflow. The oozie job command submits the workflow. It requires the oozie option that takes the Oozie server URL and the config option that takes the job.properties file.

```
oozie job -oozie http://localhost:11000/oozie -config job.properties -submit

job:14-20160525161321-oozie-test-w
```

The submit option submits the workflow to the Oozie server, and it puts the workflow in PREP status. If the workflow is successfully submitted, it returns the Oozie job ID; otherwise, it returns an error. The start command executes the Oozie workflow.

```
oozie job -oozie http://localhost:11000/oozie -config job.properties -start
```

Once the workflow is initiated, you can monitor it from the web console or the command-line interface.

The latest version of Oozie does not require `workflow.xml` to be written. It comes with the `oozie pig` command that generates `workflow.xml`. Only a complex workflow requires `workflow.xml` to be written.

> Write two Pig scripts. The first one lists all employees whose designation is "software engineer" and writes data to HDFS, and the second one counts the department-wise number of employees from the first program output. Run them manually first and write an Oozie workflow including both programs to see the advantages of the Apache Oozie workflow.

Scheduling a Pig Script

After developing a workflow, you have to decide how often you want to run it. As per specific business needs, you can run it monthly, quarterly, or yearly. Rather than running the workflow manually at chosen intervals, use the Oozie coordinator to automate the process. You can run the workflow not only on a specified date and time but also based on data availability.

To use the Oozie coordinator, you must write the `job.properties` file and `coordinator.xml` file for developing the coordinator application.

You'll now learn how to write these files and integrate them with a workflow.

Writing the job.properties File

The `job.properties` file for the coordinator is the same as the workflow `job.properties`. The only difference is that the `job.properties` file has the `oozie.coord.application.path` property instead of the `oozie.wf.application.path` property. The `oozie.coord.application.path` property must include the `hdfs` directory that stores coordinator files. The `job.properties` file is in the local file system like the workflow `job.properties` file.

Writing coordinator.xml

The file `coordinator.xml` starts with the tag `<coordinator-app>`, which has the following attributes: `start`, `end`, `frequency`, and `timezone`.

start

`start` contains the date and time when you want to run the workflow application. You can hard-code this value in `yyyy-mm-ddThh:mmZ` format. Or you can declare a parameter as `${start}` so that you can specify its value in the `job.properties` file. You can control the coordinator from the `job.properties` file, available in the local file system.

end

`end` specifies the date and time when you want to stop running the workflow application. You can hard-code this value in `yyyy-mm-ddThh:mmZ` format. Or you can declare a parameter as `${end}` so that you can specify its value in the `job.properties` file.

frequency

frequency determines how often you want to run the workflow application. It could be monthly, quarterly, or yearly or as per your requirements. Many options are available to run it as per your business requirements.

timezone

This specifies the time zone where you want to run your workflow. For example, you can specify GMT or UTC.

The following code runs the workflow application every ten minutes starting from 2016-06-02T09:00Z and ending at 2016-06-02T10:00Z in the UTC time zone.

```
<coordinator-app name="cron-coord" frequency="${coord:minutes(10)}"
start="2016-06-02T09:00Z" end="$2016-06-02T10:00Z" timezone="UTC"
                xmlns="uri:oozie:coordinator:0.2">
```

The following code runs the workflow application every ten minutes taking the start value and the end value from the job.properties file in the UTC time zone:

```
<coordinator-app name="cron-coord" frequency="${coord:minutes(10)}"
start="${start}" end="${end}" timezone="UTC"
                xmlns="uri:oozie:coordinator:0.2">
```

coord:minutes is an Oozie built-in variable. It also has coord:hours, coord:days and coord:months variables, or you can also use the cron syntax to specify how often you want to run workflow application.

Integrating with the Workflow

You can use the <workflow> tag within the <coordinator-app> tag to specify the workflow to be integrated with the coordinator. The <workflow> tag contains a child tag called <app-path> that contains the workflow application path. You can hard-code the workflow application path or you can specify a parameter that gets resolved using the job.properties variable.

The following code takes the workflow application path from a variable named workflowAppUri in the job.properties file:

```
<workflow>
            <app-path>${workflowAppUri}</app-path>
</workflow>
```

The following code contains the workflow application path /user/hdfs/examples/apps/pisg. This hdfs directory contains the workflow.xml file; required source code files such a Hive, Pig, or MapReduce; and the required JAR files.

```
<workflow>
            <app-path>/user/hdfs/examples/apps/pig</app-path>
</workflow>
```

Upload Files to HDFS

You upload the `coordinator.xml` file to the `hdfs` directory specified as the value of `oozie.coord.application.path`. You keep the `job.properties` file in the local file system.

Submitting Coordinator

You can submit the coordinator application using the Oozie command-line interface. Although you use the same command that you used for workflow submission, here you use only the coordinator `job.properties` file. If submitted successfully, the job ID is returned. For the workflow, the job ID ends with *w*, and for the coordinator application, it ends with *c*. If submission fails, it throws an error.

```
oozie job -oozie http://localhost:11000/oozie -config /path/to/job.
properties -submit

job:14-20160525161321-oozie-test-c
```

After completing the workflow exercise given in the previous section, write a coordinator application using that workflow code and make it run once daily for 30 days.

Bundle

Bundle is an Oozie application that is used for managing and launching a list of Oozie coordinators as a single application. Bundle requires the `job.properties` and `bundle.xml` files. `job.properties` is in the local file system, and the `bundle.xml` file is in HDFS. The `bundle.xml` file starts with the `<bundle-app>` tag and contains a list of `<coordinator>` tags. `<coordinator>` tags should have a child tag called `<app-path>` that contains the HDFS directory path for the coordinator application. You can submit the Bundle application in the same way as the workflow and coordinator applications using the command-line interface. The `job.properties` file must have the property `oozie.bundle.application.path`.

oozie pig Command

Oozie introduced the `oozie pig` command beginning in version 3.3.2, which simplifies the Oozie workflow submission with a Pig script.

Table 6-1 lists the options available for `oozie pig`.

Table 6-1. *oozie pig Options*

Option	Description	Example
-X <arg>	To pass parameters to Pig	-X -param input=/user/hdfs/input
-auth <arg>	To specify the authentication type	-auth SIMPLE
-config <arg>	To specify the job.properties file	-config job.properties
-doas <arg>	To specify to impersonate user	-doas oozie
-file <arg>	To specify the Pig script file	-file /path/to/pigscriptfile
-oozie <arg>	To specify the Oozie server URL	-oozie http://localhost:11000/oozie

The following command submits the dumpemp.pig script using the oozie pig command. There is no need to upload the Pig script file to HDFS because the following command automatically generates workflow.xml that you can view using the Oozie web console:

```
oozie pig -oozie http://localhost:8080/oozie -file /path/to/ dumpemp.pig
-config job.properties
```

You can pass values to parameters using the -X option.

The following code passes values to the input path and the output path parameters INPATH and OUTPATH:

```
oozie pig -oozie http://localhost:8080/oozie -file /path/to/ dumpemp.
pig -config job.properties -X -param INPATH=/user/hdfs/data/movies -param
OUTPATH==/user/hdfs/out/movies
```

When you have to pass many parameter values, you can use the paramfile option just as in Apache Pig. All you need to do is define all the values in a file and then specify the file path.

The following code uses the –param file option:

```
Cat paramvalues
INPATH=/user/hdfs/data/movies
OUTPATH==/user/hdfs/out/movies

oozie pig -oozie http://localhost:8080/oozie -file /path/to/ dumpmovies.pig
-config job.properties -X -paramfile /path/to/paramvalues
```

Command-Line Interface

Oozie provides a command-line interface that includes commands for job management, admin operations, and specific commands for each technology such as Hive and Pig as of now.

The command-line interface internally uses the REST API to interact with the Oozie server.

I'll introduce you to some of the most used commands in the Oozie CLI.

Job Submitting, Running, and Suspending

Oozie provides job command options for submitting, running, and suspending jobs.

Use the submit option to submit an Oozie job that is in PREP status. Use the run option to run an Oozie job and the suspend option to suspend a job (the job is then assigned suspend status).

The following command suspends an existing workflow:

```
oozie job -oozie http://cluster10-1:11000/oozie -suspend
0000014-160411112524820-oozie-oozi-W
```

To execute the -oozie option, the Oozie HTTP URL that is the value of the oozie. base.url property value in oozie-site.xml is mandatory.

Killing Job

Use the Oozie command-line interface option –kill to abort an Oozie job by specifying the job ID.

The following command kills the existing workflow job:

```
oozie job -oozie http://cluster10-1:11000/oozie -kill
0000014-160411112524820-oozie-oozi-W
```

Retrieving Logs

You can retrieve server logs for a specific Oozie job with its job ID using the logs option.

The following command retrieves logs for workflow job
0000014-160411112524820-oozie-oozi-W:

```
oozie job -oozie http://cluster10-1:11000/oozie -logs
0000014-160411112524820-oozie-oozi-W
```

Information About a Job

You can use the -info option to get more information about a job, such as a workflow application path, job status, and external job ID.

The following command retrieves more information about a workflow job.

The following output contains more information about a workflow job such as the username submitted, the time created, the time started, the time ended, and so on:

```
oozie job -oozie http://cluster10-1:11000/oozie -info
0000014-160411112524820-oozie-oozi-W
.------------------------------------------------------------------------------
-------------------------------------------------------------------------------
Workflow Name :  map-reduce-wf
App Path      :  hdfs://localhost:9000/user/hdfs/examples/apps/pig
Status        :  SUCCEEDED
Run           :  0
User          :  hdfs
Group         :  users
Created       :  2016-05-26 10:01 +0000
Started       :  2016-05-26 10:01 +0000
Ended         :  2016-05-26 05:01 +0000
Actions
.------------------------------------------------------------------------------
-------------------------------------------------------------------------------
Action Name            Type        Status      Transition   External Id
External Status   Error Code    Start                 End
.------------------------------------------------------------------------------
-------------------------------------------------------------------------------
hadoop1                map-reduce  OK          end
job_200904281535_0254  SUCCEEDED        -                   2009-05-26 05:01 +0000
2009-05-26 05:01 +0000
.------------------------------------------------------------------------------
-------------------------------------------------------------------------------
```

Oozie User Interface

The Oozie server comes with a user interface. oozie-site.xml contains the UI URL as a value of the oozie.base.url property. The default port is 11000. The Oozie user interface provides the status of the workflow, coordinator, and bundle jobs. Apart from the status, it also provides job logs so that you can troubleshoot failed jobs.

The Oozie user interface provides three tabs: Workflow Jobs, Coordinator Jobs, and Bundle Jobs. The tab gives more information about the jobs. Each tab has subtabs to view the status of running jobs, completed jobs, custom jobs, and all jobs. Custom jobs help filter jobs based on the status. For example, to check killed jobs, you will use a filter such as status=killed in the Custom Filter subtab.

Figure 6-2 displays workflow jobs and their information.

Figure 6-2. *Workflow jobs*

Click the Oozie job ID for more information about the job. It displays job definition, its logs, and the list of actions it executed. You can click any action to determine its external Hadoop job ID and resource manager URL to determine its progress. The Action page displays the Oozie error code if the Oozie job has failed and helps troubleshoot failed jobs.

Figure 6-3 displays more information about a workflow job.

Figure 6-3. *More info about a job*

Developing Oozie Applications Using Hue

Hue provides a user interface called the Oozie Editor to develop Oozie applications. It simplifies Oozie application development such that even a layperson can develop Oozie applications. It supports workflows, coordinators, and bundle applications. Visit `http://gethue.com/category/oozie/` for more information.

Summary

In this chapter, you learned about the workflow and scheduler engine called Apache Oozie.

These are some of important things you learned in this chapter:

- How to manage dependency among Hadoop ecosystem jobs using workflow

- How to schedule workflows using the coordinator

- That Bundle is a collection of workflow and coordinator jobs

- How to submit Pig Latin code using the `oozie pig` command

- How to use the command-line interface of Apache Oozie

- How to use the Oozie user interface for monitoring and managing Oozie jobs

For more information about Apache Oozie, you can read *Apache Oozie Essentials* by Jagat Singh.

CHAPTER 7

HCatalog

As we discussed in Chapter 1, Apache Hive is a scalable data warehousing technology built on Apache Hadoop. Hive comes with a metastore service that maintains metadata so that users can run any number of queries on already created tables. However, data processing technologies such as MapReduce and Pig do not have a built-in metadata service, so users must define a schema each time they want to run a query.

HCatalog is a table and storage management layer for Hadoop that addresses this problem by exposing Hive metadata to technologies such as MapReduce and Pig. HCatalog provides technology-independent tables in Hadoop so that users can easily read and write data without mentioning or maintaining the data schema. Tables added in Hive technology are reflected in Hcatalog, and vice versa. HCatalog internally uses the Hive metastore to serve user requests. Currently, HCatalog supports requests from Hive, Pig, and MapReduce technologies.

Initially HCatalog was an independent project at the Apache Software Foundation; now it is part of the Apache Hive project.

Features of HCatalog

The following are the salient features of HCatalog:

- HCatalog hides the data location and underlying data formats. It supports formats such as RCFile, CSV, JSON, Parquet, and SequenceFile. It also supports custom data formats.

- MapReduce, streaming, and Pig use the HCatalog API to process tables.

- In addition to the API, HCatalog provides a command-line interface (CLI) and WebHCatalog service.

- The HCatalog CLI supports DDL commands such as create, alter, and drop.

- WebHCatalog allows users to submit HTTP requests to HCatalog. WebHCatalog also helps integrate other tools with HCatalog. The data warehousing tool Teradata and the big data analytics tool Asterdata use WebHCatalog to extract data from the Hadoop ecosystem.

© Balaswamy Vaddeman 2016
B. Vaddeman, *Beginning Apache Pig*, DOI 10.1007/978-1-4842-2337-6_7

- HCatalog enables interoperability among tools such as Hive, Pig, and MapReduce. The Hive output can be made available to Pig and MapReduce for processing. Similarly, the Pig and MapReduce outputs can be made available to other tools.

- MapReduce uses the HCatInputFormat and HCatOutputFormat classes for reading and writing data from HCatalog. Pig Latin uses HCatLoader and HCatStorer for reading and writing data from HCatalog.

Figure 7-1 shows the architecture of HCatalog.

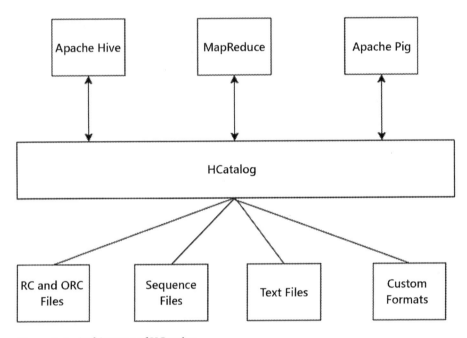

Figure 7-1. *Architecture of HCatalog*

Command-Line Interface

Just as in Hive, HCatalog comes with a command-line interface.

Use the hcat command to get to the hcatalog prompt, and use the hcat -e command to run commands.

```
hdfs@cluster10-1:~>hcat
hcat>
```

show Command

The functionality of the show command is the same as in Hive. It displays available tables, databases, and functions.

The following command displays available tables:

```
hcat -e "show tables"
OK
employee
department
Time taken: 2.305 seconds
```

Use show databases to display the databases, and use show functions to display the functions. Tables are displayed from the default database. With the use command, you can switch databases.

The following command displays tables from a user-created database called test:

```
hcat -e "use test;show tables;"
OK
Time taken: 2.069 seconds
OK
Dummy
Test
Time taken: 0.659 seconds
```

Just as in Hive, the describe command displays the schema of a table.

The following command displays the schema of table numbers:

```
hcat -e "describe numbers"
OK
num                      string
Time taken: 2.656 seconds
```

The HCatalog CLI supports only the Data Definition Language (DDL) of SQL; it does not support others such as the Data Manipulation Language and the Data Control Language.

Data Definition Language Commands

HCatalog supports only DDL commands such as the create, alter, and drop commands, and HCatalog throws an "operation not supported" error if you use other commands.

The select following command throws an error because it is not supported.

```
hcat -e "select * from numbers"
FAILED: SemanticException Operation not supported
```

create Statement

You can create tables, views, indexes, and functions using the create statement. It does not support the create table as select command of Apache Hive as it launches Hadoop jobs.

The following command creates a table named dummy with a column named test:

```
hcat -e "create table dummy (test string)"
OK
Time taken: 2.438 seconds
```

You can also create a table using another table schema.

The following command creates a new table called numbersnew using the schema of the existing table called numbers:

```
hcat -e 'create table numbernew like numbers'
OK
Time taken: 2.472 seconds
```

drop Statement

You can delete tables, views, indexes, and functions using the drop statement.

The following command drops the table named dummy:

```
hcat -e "drop table dummy"
OK
Time taken: 2.347 seconds
```

alter Statement

Use the alter statement to change the schema of a table or a view.

The following hcatalog command renames a table:

```
hcat -e "ALTER TABLE employee RENAME TO employee_new"
OK
Time taken: 2.347 seconds
```

dfs and set Commands

The HCatalog CLI also supports the dfs and set commands. dfs allows users to interact with HDFS, and set allows users to set a value of a property.

The following hcat command lists the /user/hdfs directory contents:

```
hcat -e "dfs -ls /user/hdfs "
Found 5 items
drwx------   - hdfs hdfs          0 2016-04-14 14:00 .Trash
```

```
drwxr-xr-x    - hdfs hdfs         0 2016-03-09 12:02 .hiveJars
drwx------    - hdfs hdfs         0 2016-04-14 03:34 .staging
drwx------    - hdfs hdfs         0 2016-04-14 00:50 jobstatus
drwxr-xr-x    - hdfs hdfs         0 2016-03-09 12:29 joinnumbers
```

WebHCatalog

WebHCatalog, previously called the Templeton service, is a built-in REST API in HCatalog that allows developers to submit HTTP requests to Hive, Pig, and MapReduce.

The WebHCatalog URL format is `http://ipaddress:portnumber/templeton/v1/resource`.

- `ipaddress` is the destination where the HCatalog server runs. Some distributions come with multiple HCatalog servers.

- `portnumber` is the destination where its port is listening. The default port number is 50111. To identify the value of `templeton.port`, visit `wenhcat-site.xml`.

- `resource` could be `hive`, `pig`, or `mapreduce`.

This URL will also support additional parameters that will be passed to the HTTP POST method for further processing.

Let's now discuss how to submit requests to Pig using WebHCatalog.

The WebHCatalog base URL is `http://ipaddress:port/templeton/v1/pig`.

Table 7-1 lists parameters you can pass to the Pig script in WebHCatalog.

Table 7-1. *WebHCatalog Parameters*

Parameter	Description	Comments
execute	Runs Pig Latin code	Either execute or the file is required
file	Pig Latin script file name	Either execute or the file is required
arg	Used to provide an argument	None
files	Comma-separated files to be copied to MapReduce cluster	None
statusdir	Directory where output status is written	hdfs directory
callback	URL to be called after job completion	None

Now you will learn a few ways to use WebHCatalog.

Executing Pig Latin Code

Now you will learn how to run Pig Latin code using WebHCatalog. For example, from a list of employees, if you want to get five employee records, you will write Pig Latin code like this:

```
employees = load 'employees';
employees5= limit employees  5;
store employees5 into 'employees5';
```

To run the previous code using WebHCatalog, you must run the `curl` command with the Pig base URL. The complete Pig Latin code will be executed. The `-s` option shows an error if it fails. The `-d` option takes extra data as parameters.

The following `curl` command displays five employee records by executing Pig Latin code embedded:

```
curl -s -d execute= 'employees = load 'employees'; employees5 = limit
employees 5;store employees into 'employees5' ' http://10.20.30.1:50111/
templeton/v1/pig?user.name=hdfs'
```

`user.name` is mandatory. It throws an error if the script is not submitted successfully or returns a job ID if successfully submitted. The job status can be viewed in the Resource Manager UI. This functionality is useful for small code with only a few lines.

Running a Pig Latin Script from a File

When you have to run several lines of code, you need to use the `file` option.

Here is step-by-step guide to run Pig Latin scripts from a file in WebHCatalog:

1. Write a Pig Latin script to a file.

   The following code retrieves five employee records and stores them in a directory:

   ```
   hdfs@cluster10-1:~> Cat employees5.pig
   employees = load ' employees ';
   employees5 = limit employees 5;
   store employees5 into ' employees5 ';
   ```

2. Upload the file to the `hdfs` directory.

   hdfs@cluster10-1:~> hdfs dfs -put employees5.pig **/user/hdfs**

3. Run the `curl` command.

   ```
   curl -s -d file=/user/hdfs/employees5.pig -d arg=-v 'http://
   10.20.30.1:50111/templeton/v1/pig?user.name=hdfs'
   ```

HCatLoader Example

A Pig Latin script with an HCatLoader function can also be run using WebHCatalog. For example, the following code displays five employee records:

```
employees = LOAD 'employees' USING org.apache.hive.hcatalog.pig.
HCatLoader();
employees5 = limit employees 5;
dump employees5;
```

You can run the previous program with or without the useHCatalog option. In older versions, it was mandatory that you use the useHCatalog option. In the latest versions, it is not required.

1. Write the previous code in a file.

2. Upload the previous code file to hdfs.

    ```
    hdfs dfs -put  dumpemphcatloader.pig /user/hdfs
    ```

3. Run the curl command by specifying useHCatalog as the arg parameter.

    The following command uses the useHCatalog option to run Pig Latin code using WebHCatalog:

    ```
    curl -s -d file= dumpemphcatloader.pig -d arg=-useHCatalog
    'http:// 10.20.30.1:50111/templeton/v1/pig?user.name=hdfs'
    ```

Writing the Job Status to a Directory

Use the following code to write the job status to a directory using the statusdir parameter:

```
curl -s  -d file=dumpnumhcatloader.pig -d statusdir=jobstatus
'http:// 10.20.30.1:50111/templeton/v1/pig?user.name=hdfs'
```

A new directory is created in hdfs with the subdirectories exit, stderr, and stdout.

```
   hdfs dfs -ls /user/hdfs/jobstatus
Found 3 items
-rw-r--r--   3 hdfs hdfs          2 2016-04-14 00:50 jobstatus/exit
-rw-r--r--   3 hdfs hdfs        160 2016-04-14 00:50 jobstatus/stderr
-rw-r--r--   3 hdfs hdfs          0 2016-04-14 00:50 jobstatus/stdout
```

If the curl command fails, run it in verbose mode using the -v option. WebHCatalog runs both scripts that contain HCatLoader/HCatStorer and other storage functions such as PigStorage.

HCatLoader and HCatStorer

You can use HCatLoader and HCatStorer to process Hive tables using Pig Latin. HCatLoader and HCatStorer are two functions used for reading and writing data. These two internally use the HcatInputFormat and HcatOutputFormat classes.

Reading Data from HCatalog

HCatLoader reads the data from the table specified in the load statement.

```
movies = LOAD 'movies' USING org.apache.hive.hcatalog.pig.HCatLoader();
```

If you are using a nondefault database, you must specify your input as dbname. tablename.

The following Pig Latin code displays ten employee records from a table called employees:

```
employees = LOAD 'employees' USING org.apache.hive.hcatalog.pig.HCatLoader();
employees10 = limit employees 10;
dump employees10;
```

The previous code reads data from a Hive table called employees. Pig Latin cannot read the location of the table and its data format. This Pig Latin code is dynamic and works even if the table storage location and format are changed. The schema definition is optional as it can be accessed from the existing HCatalog table schema.

To retrieve a specific portion of data from a partitioned table, write the Filter statement immediately after the load statement. A Filter statement can have multiple conditions. Conditions on the partition column are sent to HCatalog, and conditions on a nonpartitioned column are processed by Pig Latin.

The following code retrieves employee records whose date of joining is 01-01-2010 from department number 100:

```
Employees = LOAD 'employees' USING org.apache.hive.hcatalog.pig.HCatLoader();
filterdoj= filter employees by dateoj='01-01-2010' dept='100';
```

The previous code contains the partition column (date) and the nonpartition column (year). The date condition is passed to HCatalog so that it only retrieves data stored on 01-01-2010. The data for department number 100 is filtered using Pig Latin code.

Writing Data to HCatalog

To write data to HCatalog using HCatStorer, specify the table name after the into operator. The table name must be present; otherwise, it will throw a "table not found" error. If you are using a nondefault database, you must specify your input as dbname.tablename.

```
store ratings2010 into 'ratings2010' using org.apache.hcatalog.pig.HCatStorer();
```

You can write data to specific partition; you need to specify the partition column in HCatStorer in single quotes.

```
store ratings2010 into 'ratings2010' using org.apache.hcatalog.pig.
HCatStorer('year=2010');
```

If you have multiple partitions, you do not need to specify partitions in HCatStorer. However, the partition column must be included in the alias you are writing.

The following code writes the ratings1011 relation data using HCatStorer:

```
store rating1011 into 'ratings1011' using org.apache.hcatalog.pig.HCatStorer()
```

If multiple years of data is available in an alias called years1011, it will write all years of data to its respective year partition in the target table. The alias rating1011 should have the partition column year.

The table must be available in HCatalog; otherwise, the previous statement fails. Once data is written to the Hive table, it can be used by Hive, MapReduce, and again by Pig, if required.

Running Code

You need to start the Grunt shell with the useHCatalog option to run Pig Latin code with the HCatLoader function. The useHCatalog option will help Pig to locate the required JARs for processing a table.

```
hdfs@cluster10-1:~> pig -useHCatalog
```

You can run the previous Pig Latin code from a file using the -useHCatalog option.

```
pig -useHCatalog -f dumpmovies10.pig
```

You can even run a Pig Latin script without the -useHCatalog option, but you need to have HADOOP_HOME, HIVE_HOME, HCAT_HOME, PIG_CLASSPATH, and PIG_OPTS set.

HADOOP_HOME, HIVE_HOME, and HCAT_HOME must point to their home directories. PIG_CLASSPATH should have the required Hive, HCatalog, and Hadoop JARs.

In the following example, you must replace * with the actual version number.

```
export PIG_CLASSPATH=$HCAT_HOME/share/hcatalog/hcatalog-core*.jar:\
$HCAT_HOME/share/hcatalog/hcatalog-pig-adapter*.jar:\
$HIVE_HOME/lib/hive-metastore-*.jar:$HIVE_HOME/lib/libthrift-*.jar:\
$HIVE_HOME/lib/hive-exec-*.jar:$HIVE_HOME/lib/libfb303-*.jar:\
$HIVE_HOME/lib/jdo2-api-*-ec.jar:$HIVE_HOME/conf:$HADOOP_HOME/conf:\
$HIVE_HOME/lib/slf4j-api-*.jar
```

PIG_OPTS should point to the Hive metastore URI.

```
export PIG_OPTS=-Dhive.metastore.uris=<value-from-hive-site.xml>
```

Data Type Mapping

HCatalog follows the Hive data types, and they are different from the Pig data types. Hence, Pig interprets HCatalog data types using its own data types.

Table 7-2 maps HCatalog data types to Pig data types.

Table 7-2. *HCatalog Data Types to Pig Data Types*

HCatalog Data Type	Pig Data Type
Int	Int
TinyInt	Int
BigInt	Long
SmallInt	Int
Date	Datetime
TimeStamp	DateTime
Decimal	BigDecimal
Float	Float
Double	Double
Char	chararray
Varchar	chararray
String	Chararray
Boolean	Boolean
Binary	Bytearray
Map	Map
Struct	Tuple
List	Bag

Before writing data into the HCatalog table, you must check current alias data types against the target table data type. If they cannot be mapped, the Pig Latin script throws an error.

For example, if you have the int data type in the Pig Latin alias and try to write that column data into the string data type of the HCatalog table, it will throw the following error:

```
Pig 'int' type in column 0(0-based) cannot map to HCat 'STRING'type.  Target
filed must be of HCat type {INT or BIGINT or TINYINT or SMALLINT}
```

However, many new data types have been added to both Hive and Pig. Some new data types might not have a direct mapping in Pig Latin. When there is no mapping in Pig, HCatalog checks for the range of the target data type. If the range does not match, it will insert null values by default. However, you can display an error rather than inserting null values.

Use -onOutOfRangeValue Throw in the HCatStorer function to throw an error:

```
store data into 'movies10' using org.apache.hive.hcatalog.pig.
HCatStorer('','','-onOutOfRangeValue Throw');
```

Summary

In this chapter, you learned about the table and storage management tool HCatalog.
  Here are some of the important things you learned about HCatalog:

- Data formats supported by HCatalog

- How it achieves interoperability among Hive, Pig, and MapReduce

- How to use DDL commands in the command-line interface of HCatalog

- How to access HDFS and set property values

- How to submit Pig Latin code using WebHCatalog

- How to read data using HCatLoader and how to write data using HCatStorer

- The data type mappings between Pig and HCatalog

CHAPTER 8

▩ ▩ ▩

Pig Latin in Hue

Every technology in the Hadoop ecosystem comes with a command-line interface that enhances the user experience with the technology. The Hadoop ecosystem is replete with technologies, and it is impossible to remember all of the commands, which are also case sensitive. Hue (which stands for Hadoop User Experience) alleviates this problem by providing web interfaces for most of the technologies in the Hadoop ecosystem.

Hue aggregates a set of web applications related to the Hadoop ecosystem and provides a common graphical user interface that makes using Hadoop easy. Some of commonly used web applications are a File Browser utility and a Job Browser utility. Hue provides the File Browser utility to manage the Hadoop file system and the Job Browser utility to manage Hadoop jobs. An integrated user interface is also available for tools such as Hive, Pig, Sqoop, Oozie, Impala, and HBase. Hue is available on distributions such as Cloudera, Hortonworks, and MapR.

Hue sits in the middle of the user and the Hadoop ecosystem. It serves user requests using graphical user interface for its integrated web applications. For example, a user can submit Hive jobs using the Hive web application (Beeswax) in Hue without access to the Hive command line. Similarly, you can submit Oozie jobs using the Oozie web application in Hue. The process is further simplified because the Oozie script is generated through the Oozie dashboard.

Hue is useful for security reasons to restrict Hadoop users from web applications so that users need to know less about the Hadoop cluster. It further simplifies Hadoop job management for users by providing a graphical user interface.

This chapter will discuss the Pig module, File Browser utility, and the Job Browser utility.

Pig Module

The Pig module in Hue improves the user experience while submitting Pig jobs. You can save Pig Latin scripts and view previously run scripts and logs of completed jobs. You can also view the progress of jobs that are running. The Pig module in Hue is divided into two functionalities: My Scripts and Query History.

© Balaswamy Vaddeman 2016
B. Vaddeman, *Beginning Apache Pig*, DOI 10.1007/978-1-4842-2337-6_8

Figure 8-1 shows the Pig Editor in Hue.

Figure 8-1. *The Pig Editor*

My Scripts

My Scripts allows users to write, save, modify, and run Pig Latin scripts.
Click New Scripts to write a new Pig Latin script.
Select a Pig Latin script from My Scripts to modify or execute it.
Figure 8-2 displays the Pig Editor with Pig Latin code.

Figure 8-2. *Pig Editor with code*

Clicking the New Script link provides a large text area to write Pig Latin scripts and a text box to set the title for the script to be written. In addition, it provides the following buttons: Save, Execute, Explain, and Syntax check.

> *Save*: Click Save to save the script. Enter the name for the script to be saved in the Title text box.

> *Execute*: Click Execute to run the Pig Latin code you have written. It functions like the run command.

> *Explain*: Click Explain to display the explain plan of the script. It is equivalent to the Pig Latin explain operator.

> *Syntax check*: Click "Syntax check" to display syntax errors if any are found.

Once a job is submitted, Kill appears. Click Kill to kill a job that is running. Once a job is completed, the Logs link appears; click it to view a log of jobs.

Pig Helper

Pig Helper in the Pig module displays the syntax of all the operators. When you select an operator, the syntax of that operator appears in the text area. In the text area, you can modify the syntax to suit your requirement.

For example, if you do not remember the HCatLoader syntax, you can select the HCatalog syntax from HCatalog drop-down box in the Pig Helper. The syntax appears in the text area, and you can modify the table name to choose the table you want to view. Thus, you don't have to remember the syntax of Pig Latin operators.

Figure 8-3 shows the Pig Helper options.

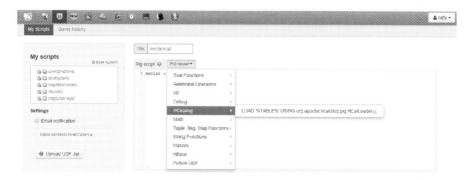

Figure 8-3. *Pig Helper options*

Auto-suggestion

The auto-suggestion option is available in the Hue UI. If you enter the starting letter of a Pig Latin operator and click Ctrl+Enter, a list of Pig Latin operators appears on the screen. Auto-suggestion is also available in the path section of the load or store operator. Click Ctrl+Enter and a list of folder names or file names appears. The auto-suggestion option is particularly useful while searching for case-sensitive operators.

Figure 8-4 displays the auto-suggestion options.

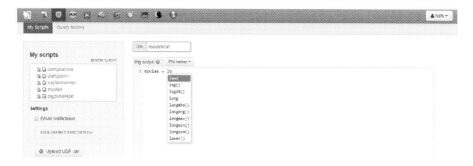

Figure 8-4. *The auto-suggestion options*

117

UDF Usage in Script

User-defined functions (UDFs) can be quite easily used in Pig Latin scripts. Click the Upload UDF Jar button to upload a JAR file in the Pig module of Hue. Once the UDF is uploaded, it appears in the User Defined Functions drop-down menu. Click the UDF JAR link to automatically generate the `Register` command for the chosen UDF.

Figure 8-5 displays the User-Defined Functions drop-down menu.

Figure 8-5. *User-Defined Functions list*

Query History

Query History displays a list of the last-run Pig scripts. The list displays the following columns: Date, Pig Script, and Status. It also provides a Delete option. The Date column shows the date and time when a Pig script was run. The Pig Script column shows the name of the script was run. Click the script name to view the output generated in that run. Click the Execute button to rerun the script. Click the Delete button to remove the script run and its output.

Figure 8-6 displays the Query History area.

Figure 8-6. *Query History area*

File Browser

Users do not need to have terminal access to interact with the Hadoop Distributed File System. The File Browser utility provides the HDFS Shell Guide feature in Hue. Users do not need to remember HDFS shell commands. Even a layperson can explore HDFS using the File Browser utility in Hue (see Figure 8-7).

Figure 8-7. *File Browser utility*

File Browser allows users to create, modify, and delete files and directories on HDFS.

1. To create a file, select New ➤ File.

2. Enter a file name in the Create File box (see Figure 8-8).

3. Click Submit.

Figure 8-8. *Create File box*

To create a new directory, follow these steps:

1. Select New ➤ Directory.

2. Enter a directory name in the Create Directory box.

3. Click Submit. The new directory created in Hue is equivalent to the `hdfs dfs -mkdir` command (see Figure 8-9).

Figure 8-9. *Directory name*

119

To upload files to HDFS using File Browser, follow these steps:

1. Click Upload ➤ Files.

2. Click again to upload a file and select the files to open.

3. Select Upload ➤ Zip file to upload the ZIP files (see Figure 8-10).

Figure 8-10. *Unzipping files*

This functionality is the same as the `copyFromLocal` command in the HDFS Shell Guide.

To download HDFS files, follow this step:

1. Click the Download button (see Figure 8-11) after selecting the files in the File Browser.

Figure 8-11. *Downloading files*

This functionality is the same as the `copyToLocal` command of the HDFS Shell Guide.

The File Browser allows users to change the permissions of files and directories. You can even change the owner and groups.

1. Select the check box next to File/Directory and click the Change Permissions button.

2. In the Permissions page, you can choose the read, write, or execute permissions and click Submit (see Figure 8-12).

120

Figure 8-12. *Permissions*

This function is equivalent to the hdfs dfs -chmod command.
You can search files and directories in HDFS.

1. Enter a file or directory name in the search box (see Figure 8-13).

Figure 8-13. *Search feature*

A list of all matching files and directories appears.
You can also copy/move files or directories within the Hadoop distributed file system.

Job Browser

You do not have to access the Job Tracker or Resource Manager to view information on jobs. The Job Browser in Hue allows you to view different types of jobs such as running, failed, killed, and finished. The Job Browser displays two text boxes: Username and Text. In the Username text box, enter the username to view jobs executed by that user. In the Text box, enter the job name to search for the job. Matching jobs are displayed. Jobs can be searched by job ID or by job name. The status of the job is also displayed (see Figure 8-14).

Figure 8-14. *Job status*

Click the job ID to view more information about the job.

Click Logs to view the job logs.

Click the Succeeded button to see the jobs that completed successfully.

Click the Running or Failed or Killed button to see the jobs.

Running jobs display the following: percentage of maps completed, percentage of reducers completed, time taken for job, and queue name. The Job Browser functionality is equivalent to the mapred job command.

Summary

In this chapter, you learned about three Pig-related features of Hue.

- How to use the Pig Editor features such as how to write new scripts, how to use the Pig Helper and auto-suggestion, and how to use UDF JARs in a Pig Latin script

- How to do file operations such as creating files, deleting files, moving files, and so on, in the File Browser application of Hue

- How to perform job management in the Job Browser of Hue

CHAPTER 9

∎ ∎ ∎

Pig Latin Scripts in Apache Falcon

In this chapter, you will learn all about Apache Falcon and how to use Pig Latin scripts in Falcon. Apache Falcon is a Hadoop framework used for data lifecycle management. Its applications include data feed management, data replication from one cluster to another, and a lineage of data applications. Although developed by InMobi, it is now an Apache project.

Feed management in Apache Falcon allows users to do the following:

- Define the retention period to retain data

- Define the retry policy to handle job failures and manage late data arrival

Most enterprises use Apache Falcon to replicate data from production to the backup cluster or disaster recovery cluster.

Falcon makes use of Hadoop ecosystem tools such as Apache Oozie, Apache Pig, and core Hadoop to provide these data applications. Apache Falcon is built on top of Apache Oozie, and it works as a high-level abstraction.

Falcon contains three entities called cluster, feed, and process. The feed and process entities will have the cluster entity internally. The process entity will have the feed entity internally.

- The cluster entity defines the Hadoop cluster to be used.

- The feed entity defines the data set location and late data arrival management.

- The process entity defines the workflow and retries the workflow execution in the case of job failures.

Figure 9-1 shows the relationship among these entities.

© Balaswamy Vaddeman 2016
B. Vaddeman, *Beginning Apache Pig*, DOI 10.1007/978-1-4842-2337-6_9

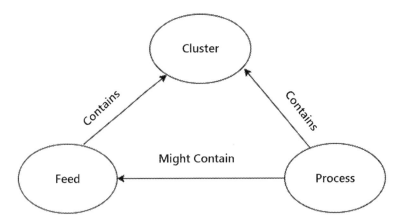

Figure 9-1. *Relationship between Falcon entities*

cluster

Now you will learn how to write the `cluster` entity. The `cluster` entity must have a unique name. `colo` defines a colocated cluster.

The following code defines the `cluster` entity and its colocated cluster:

```
<cluster colo="drcluster" description="" name="devcluster"
xmlns="uri:falcon:cluster:0.1"
 xmlns:xsi="http://www.w3.org/2001/XMLSchema-instance">
```

Interfaces

Falcon utilizes many services to complete a job. For example, it uses HDFS to create temporary directories and Hive to retrieve metadata and partitions. All such services must be defined as interfaces in the `cluster` entity.

Now you will learn how to write interfaces in the `cluster` entity.

The following code defines an HFTP interface to read data from the remote cluster:

```
<interface type="readonly" endpoint="hftp://<hostname>:50010" version="0.20.2" />
```

In the following code, the `write` interface is the value of the property `fs.defaultFS` in `core-site.xml`. It is used for writing data to the staging directories.

```
<interface type="write" endpoint="hdfs://<hostname>:8020" version="0.20.2" />
```

The name node's hostname must be used to replace `<hostname>` in the previous code.

In the following code, the execute interface represents the value of the property `mapreduce.jobtracker.address`. It submits jobs. Use the job tracker hostname in the endpoint URL.

```
<interface type="execute" endpoint="<hostname>:8021" version="0.20.2" />
```

The following code defines the Oozie endpoint:

```
<interface type="workflow" endpoint="http://<hostname>:11000/oozie/"
version="4.0" />
```

Falcon depends on Oozie for job scheduling. The `workflow` interface specifies the Oozie URL. The endpoint URL must include the name of the host where Oozie is running.

In the following code, the `registry` interface specifies the metadata service of Apache Hive. It interacts with Hive partitions.

```
<interface type="registry" endpoint="thrift://<hostname>:9083" version="0.11.0" />
```

The `messaging` interface allows for messaging internally within a Falcon job. The following code defines the messaging interface:

```
<interface type="messaging" endpoint="tcp://<hostname>:61616?daemon=true"
version="5.4.6" />
```

Locations

A cluster has a list of locations defined as follows:

```
<location name="staging" path="/tmp/emp/staging" />
<location name="working" path-"/tmp/cmp/working" />
```

`location` has attributes called `name` and `path`. The `name` attribute defines the type of location. Locations are three types: staging, temp, and working. All paths point to the HDFS location. These locations must be present before executing Falcon entities and must have read/write/execute permissions for the Falcon user.

- The working directory is optional but is not the staging directory.

- Staging needs to have 777 permissions, and the working directory should also have 755 permissions.

Write the `cluster` entity and submit it to the Falcon server using the following code:

```
falcon@cluster10-1:~> falcon entity -type cluster -file prodCluster.xml -submit
```

feed

feed defines the data set. The feed XML begins with the feed tag, and it must have a unique name. The feed entity is used by the process entity.

The following code defines the feed entity named ratings:

```
<feed description="ratingsfeed" name="ratings" xmlns="uri:falcon:feed:0.1"
xmlns:xsi="http://www.w3.org/2001/XMLSchema-instance">
```

Feed Types

There are two types of feed: the file system that refers to the hdfs directory path or the Hive table.

File System

The file system feed defines paths related to HDFS. It uses locations and location tags. The <locations> tag contains <location> tags that refer to the data location, the stats location, and the meta locations.

The following code defines the data, stats, and meta locations.

```
<locations>
<location type="data" path="/data/mvoies" />
 <location type="stats" path="/data/falcon/stats " />
 <location type="meta" path="/data/falcon/meta" />
</locations>
```

Table

You can specify a Hive table as a type of feed. You can also specify partitions. Here's the syntax:

```
catalog:$database-name:$table-name#partition-key=partition-value
```

The following code defines a ratings table in the emp_db database with the partition key as the date:

```
<table uri="catalog:emp_db:ratings#date=${YEAR}-${MONTH}-${DAY}" />
```

Frequency

The <frequency> tag specifies the frequency of the data feed. You can use minutes, hours, days, and months as frequency variables.

The following tag specifies the feed to be generated once every day:

```
<frequency>days(1)</frequency>
```

Late Arrival

The `<late-arrival>` tag specifies the grace period for the late arrival of the feed using the `cut-off` attribute.

The following code specifies that the feed can be a maximum of five minutes late:

```
<late-arrival cut-off='minutes(5)'/>
```

Cluster

`feed` uses the `cluster` entity that you have already submitted. Along with `cluster`, you specify the validity and the retention period. `validity` specifies the start date and the end date for a feed.

Use the `retention` tag to specify a period of retention. `retention` has two attributes, namely, `limit` and `action`.

- The `limit` attribute specifies a period of time for the feed to be retained.

- The `action` attribute specifies an action that must be performed after the end date.

The cluster in the following code is valid for one year. After one year, the feed will be retained for three months and deleted after three months.

```
<cluster name="prodcluster">
          <validity start="2016-06-20T00:00Z" end="2017-06-19T00:00Z"/>
          <retention limit="months(3)" action="delete"/>
 </cluster>
```

After writing the feed entity, submit it to the Falcon server and schedule it to get it into a running state.

The following command submits the feed:

```
falcon@cluster10-1:~> falcon entity -type feed -file ratingsfeed.xml -submit
```

The following code schedules the last-submitted feed:

```
falcon@cluster10-1:~> falcon entity -type feed -name ratingsfeed -schedule
```

process

The process entity defines the process to be used for executing a workflow. process can be defined with or without the existing the feed entity. The process entity talks to the cluster entity but may not talk to the feed entity. The process entity must have a unique name.

You define the process entity using the <process> tag as follows:

```
<process name="ratingprocess">
</process>
```

The following are some important attributes to be defined for process.

cluster

The cluster attribute defines the Hadoop cluster for executing a workflow. The <clusters> tag has one or more child <cluster> tags that will point to an existing cluster entity. Every cluster will have start time and end time defined by the <validity> tag, as shown in the following code:

```
<clusters>
        <cluster name="prodCluster">
            <validity start="2016-06-21T16:15Z" end="2017-06-21T16:15Z "/>
        </cluster>
        .
        .
        .
        <cluster name="DRCluster">
            <validity start="2016-06-21T16:15Z" end="2017-06-21T16:15Z "/>
        </cluster>

</clusters>
```

Failures

In the case of job failures and if you want to rerun the workflow, you can use the <retry> tag. The <retry> tag has attributes to specify the time interval after which the workflow must be rerun, the number of times it must be rerun, and whether it should be rerun in the case of a timeout error.

The following code defines a retry policy with a 30-minute delay for a maximum of three times. It also specifies to rerun the workflow on a timeout error.

```
<retry policy= "periodic" delay= "minutes(30)" attempts=3 onTimeout='true'/>
```

feed

Specifying feed in process is optional. You can specify feed either as input or as output to a workflow. The workflow takes the input data set from the input feed and writes to the output data set specified in the output feed.

The following code specifies the output feed:

```
<outputs>
        <output name="ratingsOut" feed="ratingfeed" instance="today(0,0)"/>
    </outputs>
```

workflow

You can define the process entity on the existing Oozie workflow or you can choose the Hive script or Pig script as a workflow engine. You must mention the engine, its version, and the script path in the HDFS directory, as shown in the following code:

```
<workflow engine="pig" version="0.15" path="/user/hdfs/apps/pig/ratings.pig"/>
```

Once you write the process entity, you submit it to the Falcon server using the submit option and schedule it to get into running mode, as shown in the following code:

```
falcon@cluster10-1:~> falcon entity -type process -file ratingsprocess.xml -submit
```

```
falcon@cluster10-1:~> falcon entity -type process -name ratingsprocess -schedule
```

CLI

A user can interact with Falcon using either the command-line interface or the Falcon web interface.

The Falcon CLI provides commands for entity management, instance management, metadata, admin, and recipe commands. CLI internally uses the REST API of Falcon to serve user requests.

Now you will learn about some important commands of CLI.

entity

The Falcon entity command provides options to manage all entities. You can submit an entity, schedule an entity, and delete an entity.

Submit

After developing an entity, you submit it to the Falcon server. The status of the entity changes to Submitted. The Type option specifies the type of entity you are submitting, such as cluster, feed, or process. The file option must refer to the file path of an XML.

The following code submits the cluster entity using its definition in prodCluster.xml:

```
falcon@cluster10-1:~> falcon entity -type cluster -file /path/to/
prodCluster.xml -submit
```

Schedule

After you submit an entity, you must schedule it so that you can run it. The feed and process entities can be scheduled. Use the name defined for the entity in XML to schedule it.

The following code schedules a feed named ratingsfeed:

```
falcon@cluster10-1:~> falcon entity -type feed -name ratingsfeed -schedule
```

Suspend

You can suspend an entity that is in running state using the suspend option. Only feed and process can be suspended. Write the name of the entity you want to suspend.

The following code suspends the feed entity named ratingsfeed:

```
falcon@cluster10-1:~> falcon entity -type feed -name ratingsfeed -suspend
```

Resume

You can resume an entity that is in suspended state. Like suspend, resume can be applied only on feed and process.

The following code resumes the feed named ratings:

```
falcon@cluster10-1:~> falcon entity -type feed -name ratingsfeed -resume
```

Apart from the previously mentioned commands, you can delete an entity and update the definition of an entity. The Falcon instance command provides options for killing, suspending, and rerunning instances also.

Web Interface

The Falcon web interface provides features for managing entities including a search option, developing entities, notifications, and mirrors.

Search

You can search an entity using a name, or you can use * to list all entities. In the output, you can select an entity and perform operations such as scheduling, suspending, resuming, and so on.

Figure 9-2 shows buttons for the entities and also the search box.

Figure 9-2. *Buttons for entities*

Create an Entity

The "Create an entity" feature of the Falcon web interface helps you create an entity by generating XML. You do not need to write XML manually. All entities can be generated with this feature.

In the user interface, you enter XML tag values and their attribute values. XML is generated as you enter the values. The XML code is displayed in the right section of the user interface. You can edit XML using Edit XML and then save it.

The saved entity in the user interface is submitted to the Falcon server. To search for a submitted entity, enter its name in the search box. In the search results, choose an entity by clicking the check box. To schedule the chosen entity, click the Schedule button.

Notifications

To view all notifications including error messages generated during entity submission, click the Notifications link.

Mirror

The Mirror feature allows users to create a Falcon entity that replicates data from one cluster to another.

Data Replication Using the Falcon Web UI

Many enterprises replicate data from the production cluster to a backup or disaster recovery cluster.

The HDFS DistCp program is a good option to copy data from one cluster to another. You may need to run DistCp multiple times and rerun failed DistCp programs.

- You can schedule DistCp using the Unix crontab to avoid running the DistCp program multiple times manually. But DistCp will not rerun failed DistCp programs.

- To avoid manually scheduling DistCp and manually rerunning failed jobs, you can use the Falcon web interface.

The Falcon web interface provides a mirror feature that can be used to mirror data from one cluster to another. The source cluster and destination cluster can be one of the following: HDFS, Microsoft Azure, or Amazon S3. In addition to the HDFS path, you can also mirror hive tables or databases.

Figure 9-3 shows the Mirror button in the Falcon web UI.

Figure 9-3. Mirror button

Create Cluster Entities

Create one `cluster` entity for the source and one `cluster` entity for the destination.

Create Mirror Job

Click Mirror in the Falcon web interface to display the page to create a mirror job. Enter a unique name for the mirror job in the Mirror Name field, as shown in Figure 9-4.

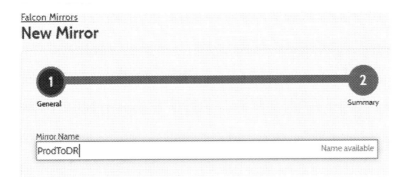

Figure 9-4. Mirror Name field

Select a mirror type between File System and Hive Catalog Storage. Select a file system to mirror data among HDFS, Azure, and S3 and select Hive (catalog storage) to mirror hive tables or databases. Select one source from HDFS, Azure, and S3.

In Figure 9-5, I have selected HDFS as the source cluster. From the drop-down menu, select the required cluster. In Figure 9-5, I have selected the `prodcluster` entity as the source cluster entity. I have entered `/data/prod/movies` as the source path.

Mirror type

| File System | HIVE(catalog Storage) |

Source ──────────────────────────── ○ Run job here

Location: ◉ HDFS ○ Azure ○ S3

prodCluster ▼

Path
/data/prod/movies

Target ──────────────────────────── ◉ Run job here

Location: ◉ HDFS ○ Azure ○ S3

DRCluster ▼

Path
/data/dr/movies

Figure 9-5. *Target settings*

Similarly for Target, I have selected HDFS as the location, DRCluster as the target cluster entity, and /data/dr/movies as the target path where the data is to be replicated. The run job here determines the cluster where the job needs to be run. Here you have to choose the target cluster to run the mirror job.

In the Validity section, specify the start date and end date for this mirror job, as shown in Figure 9-6.

Validity

| Start | Time | | | End | Time | | | |
| 06/12/2016 | 09 | 18 | AM | 06/01/2017 | 09 | 18 | AM | (GMT) Wester ▼ |

Figure 9-6. *Validity section*

In the Frequency field, enter the frequency with which the job has to be run. Here, I have specified that the job is to be run every day, as shown in Figure 9-7. In the Delay text box, specify the grace time after which a retry is to be performed and specify the number of retries.

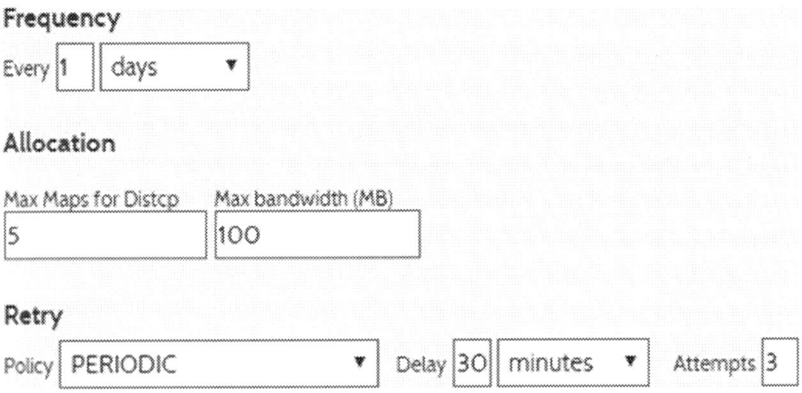
Frequency
Every 1 days
Allocation
Max Maps for Distcp
Max bandwidth (MB)
5
100
Retry
Policy PERIODIC
Delay 30 minutes
Attempts 3

Figure 9-7. *Frequency settings*

After all these details are entered, click Next and Save. The mirror job is submitted. To schedule a job, follow these steps:

1. In the search box, search for and select the mirror job.
2. Click Schedule.
3. The Oozie user interface and Resource Manager UI display the jobs running periodically to replicate data from the `prod` cluster to the `dr` cluster.

Figure 9-8 shows the mirror job submitted.

Name: prodtodr
Search
Name
Tags
Cluster
Type
Status

Figure 9-8. *Mirror job submitted*

Pig Scripts in Apache Falcon

Apache Pig can be used to process feeds. You can use `pig` *in* Apache Falcon to take care of handling late data, feed retention, and feed replication instead of writing code for them.

Falcon can be used on an individual Pig script and also in an Oozie workflow that contains a Pig script.

Oozie Workflow

You can use Oozie as a workflow engine in a `process` entity. You need to have a complete workflow application ready in HDFS including `workflow.xml`, script files, and required JARs.

As discussed earlier in this chapter, you will specify the Oozie workflow path and Oozie version in the process entity. The workflow can read data from the input feed and can write to the output feed if you specify the input feed and the output feed.

Pig Script

To choose an individual Pig script as a workflow engine in the process entity, upload the Pig script to the HDFS directory and specify the file path in the process entity using the Falcon web UI.

The following code defines the Pig Latin script as a workflow engine:

```
<workflow engine="pig" version="0.15" path="/user/hdfs/apps/pig/ratings.pig"/>
```

The read, write, and execute permissions on the Pig Latin script must be granted to the user who is running the Falcon process. The Pig script reads data from the input feed and writes output data to the output feed. Once a process entity is scheduled, the Bundle job is launched. The Bundle job starts the coordinator job, the coordinator job starts the workflow job, and the workflow job runs the Pig script.

The following Falcon process runs the embedded Pig script, /user/falcon/storeinhbase.pig, to transfer data from HDFS to HBase once a day. In case of failure, it attempts three times in five-minute intervals. It also includes a cluster entity called testCluster.

The following code uses Pig Latin code as a workflow engine:

```
<process xmlns='uri:falcon:process:0.1' name='hdfsToHbase'>
  <clusters>
    <cluster name='testCluster'>
      <validity start='2016-08-04T10:27Z' end='2016-09-04T12:27Z'/>
    </cluster>
  </clusters>
  <parallel>1</parallel>
  <order>FIFO</order>
  <frequency>days(1)</frequency>
  <timezone>GMT-05:00</timezone>
  <workflow name='pigtest' version='pig-0.12.0' engine='pig' path='/user/
falcon/storeinhbase.pig'/>
  <retry policy='periodic' delay='minutes(5)' attempts='3'/>
  <ACL owner='hdfs' group='hdfs' permission='0755'/>
</process>
```

From the command line, you must first submit the previous process and then schedule a process. The cluster must be submitted first and then the process entity.

The Pig Latin script will look like this:

```
emp = load '/path/to/employee/dataset' using PigStorage(',') as (empno:int,e
name:chararray,salary:int,deptno:int);
```

```
store emp into 'hbase://emp' using org.apache.pig.backend.hadoop.hbase.
HBaseStorage(
    'empdetails:ename empdetails:salary empdetails:deptno ');
```

This transfers data from an employee data set in HDFS to the emp table in HBase.

You can avoid writing the process entity manually; the Falcon web UI will generate the process entity XML for you.

Summary

Apache Falcon is a feed and process management system built on top of Apache Oozie. You learned the fundamentals of Apache Falcon in this chapter.

- How to define, submit, and schedule the Falcon entities called cluster, feed, and process

- How to use the Falcon CLI commands for submitting, scheduling, suspending, and resuming Falcon entities

- How the Falcon web UI can be used for creating entities and also searching for existing entities

- How to create mirror jobs using the Falcon web interface

- How to use Apache Oozie or Apache Pig as a workflow engine in the process entity

CHAPTER 10

Macros

In this chapter, you will learn how to write macros in Pig Latin.

For example, you will write code like the following to compute a department-wise employee count for all non-null departments on the employee data set:

```
dnogrp = group emp by dno;
empcount = foreach dnogrp generate group,COUNT(emp.eno) as eno;
empcount = filter empcount by group is not null;
dump empcount;
```

If you have a similar requirement to compute a salary-wise employee count or designation-wise employee count, you need to rewrite the entire previous code to change only the group column name. If you have to rewrite code every time you have a similar requirement, you cannot be productive. Also, it is difficult to maintain such redundant code.

Pig Latin provides a feature called a *macro* that enables you to write more reusable and maintainable Pig Latin code.

Macro can be written using the DEFINE operator followed by a name, input parameters, return values, and list of Pig Latin statements. The following is the macro structure and an explanation of its parts.

Structure

This is the structure:

```
DEFINE macro_name (param1, param2, ...]) RETURNS {void | rel1 [, rel2 ...]} {
  pig_latin_statements
  };
```

macro_name is the name of the macro to be defined. The macro name needs to be unique.

Four types of parameters can be given as input to a macro. The relation name can be an integer, string, and float. Return values can be void or one or more number of relation names. VOID is used if nothing is there to be returned. The relation that is specified should be there in the macro body as $relationname.

Macro Use Case

For this example, say you have identified reusable functionality with respect to your requirements and will write a macro for that. For the previously discussed requirement, the dno column-wise employee count for non-null values is eligible for reusable code.

You can write a macro as shown here:

```
DEFINE countagg(r,colname)  returns rerel{
dnogrp=group $r  by $colname;
empcount = foreach dnogrp generate group,COUNT($r.eno) as eno;
$rerel = filter empcount by group is not null;
}
```

The macro name is countagg, which takes two inputs. One is a relation, and the other is the column name. You are naming them r and colname, respectively. These are referred to using $ within the macro code. As a macro is not directly hard-coded in the emp relation and the dno column, you can use it for different relations and different columns also.

You can use a macro like the following in Pig Latin code to compute the department-wise employee count for non-null department values:

```
emp = load 'employee.csv' using PigStorage(',') as (eno:int,ename:chararray,
salary:int,dno:int);
empcount = countagg (emp,dno);
```

If you want to compute the salary-wise employee count, you can just change one of the input parameters (dno) to the salary column.

```
empcount = countagg (emp,salary);
```

Like this, a macro allows you to write more dynamic Pig Latin scripts by providing relation names and column names at the time of using the macro.

Macro Types

Macros come in two types, as specified in Figure 10-1.

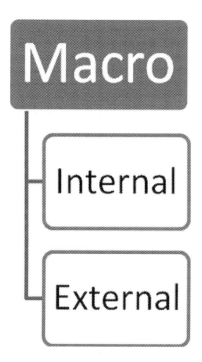

Figure 10-1. *Types of macros*

Macros can be defined internally in a Pig Latin script or externally to a Pig Latin script in a separate file.

Internal Macro

Macros can be defined internally within a Pig Latin script. You do not need to link a Pig Latin script to a macro explicitly because it is available in the same file; also, other Pig Latin scripts cannot refer to it, so an internal macro is less reusable. You can define any number of macros in a Pig Latin script.

The following Pig Latin code contains a macro called countagg within it:

```
DEFINE countagg(r,colname)  returns rerel{
dnogrp=group $r  by $colname;
empcount = foreach dnogrp generate group,COUNT($r.eno) as eno;
$rerel = filter empcount by group is not null;
}
emp = load 'employee.csv' using PigStorage(',') as (eno:int,ename:chararray,
salary:int,dno:int);
empcount = countagg (emp,dno);
dump empcount;
```

External Macro

External macros are defined in a separate file. The following are steps for writing an external macro in Pig Latin:

1. You need to write a macro in a separate file, and the file needs to be in the local file system.

   The following file called countagg.macro contains a macro called countagg:

   ```
   Cat /home/hdfs/countagg.macro
   DEFINE countagg(r,colname)  returns rerel{
   dnogrp=group $r  by $colname;
   empcount = foreach dnogrp generate group,COUNT($r.eno) as eno;
   $rerel = filter empcount by group is not null;
   }
   ```

2. You need to import a macro into a Pig Latin script using its file path. The file path can be absolute or relative. The relative path will be resolved from the current working directory in the local file system.

   The following code is an example of importing an external macro called countagg:

   ```
   IMPORT 'countagg.macro';
   ```

3. After creating a macro in a separate file and importing it, you can use it within Pig Latin code.

   The following code contains both an import statement and a macro usage statement:

```
hdfs@cluster10-1:~> Cat columnwiseempcount.pig
```

```
IMPORT 'countagg.macro';
emp = load 'employee.csv' using PigStorage(',') as
     (eno:int,ename:chararray,salary:int,dno:int);
empcount = countagg (emp,dno);       ---------macro usage
dump empcount;
```

   After these steps, you can run a Pig Latin script using a file path like the following:

```
Pig -f /path/to/columnwiseempcount.pig
```

   An external macro needs to be imported before it is used in Pig Latin, and it provides highly reusable and maintainable Pig Latin code.

dryrun

When you run a Pig Latin script that contains a macro, a macro statement in Pig Latin code will be replaced by macro code. You can view newly generated Pig Latin code using the dryrun feature.

The following code is an example of a dryrun of a Pig Latin script.

```
Pig -dryrun -f /path/to/columnwiseempcount.pig
```

When you dryrun Pig Latin code, it generates a file with the name piglatinscriptfilename.expanded. You can view a complete new Pig Latin code in that file.

```
hdfs@cluster10-1:~> cat columnwiseempcount.pig.expanded
emp = load 'employee.csv' USING PigStorage(',') as (eno:int,
ename:chararray, salary:int, dno:int);
macro_countagg_dnogrp_0 = group emp by (salary);
empcount = foreach macro_countagg_dnogrp_0 generate group, COUNT(emp.(eno)) AS eno;
empcount = filter empcount BY (group IS not null);
dump empcount;
```

Macro Chaining

When you have multiple macros, you can have one macro refer to another macro as long as it is not a recursive call. In the previous example, you wrote a macro file to compute one column-wise employee count and removed non-null values from the group column. Now you will see how to write a new macro for filtering the non-null values of a column and refer to it from the countagg macro.

Say you need to filter non-null values from a column of a relation, so you need two inputs; one is a relation, and another is a column. As it returns a new relation after removing non-null values, you have to declare a relation as the return value.

The following code contains the complete code for a new macro:

```
hdfs@cluster10-1:~> cat filternotnull.macro
DEFINE filternotnull(r1,colname1)  returns rerel{
$rerel = filter $r1 by $colname1 is not null;
};
```

You need to make two changes in the countagg macro file:

1. You need to add an import statement that contains the previously created macro file path.

2. You need to replace the earlier filter's not-null functionality with the macro call.

The following code is the new code for the countagg macro:

```
hdfs@cluster10-1:~> cat countagg.macro
IMPORT 'filternotnull.macro';
DEFINE countagg(r,colname)  returns rerel1{
dnogrp=group $r  by $colname;
empcount = foreach dnogrp generate group as group1,COUNT($r.eno) as eno;
$rerel1 = filternotnull(empcount,group1);
};
```

The Pig Latin script remains the same because the second macro is referred to only from the first macro.

The following code is the final Pig Latin script that contains macro chaining:

```
hdfs@cluster10-1:~> cat importmacro.pig
import 'countagg.macro';
emp = load 'employee.csv' using PigStorage(',') as (eno:int,ename:chararray,
salary:int,dno:int);
empcount = countagg (emp,dno);
dump empcount;
```

Macro Rules

Macros need to follow some rules, as covered next.

Define Before Usage

A macro needs to be defined before its first usage.

The following code is correct:

```
DEFINE macroex() RETURNS void {
..
..
 };
.

.
Rel =  macroex();
```

The following code is not correct:

```
..
..
Rel =  macroex();

DEFINE macroex() RETURNS void {
..
..
 };
```

CHAPTER 10 ▨ MACROS

Valid Macro Chaining

Macro chaining that is one macro referring to another macro is fine if it is not recursive. Recursive calls are not allowed in macros. This applies for both a single macro and multiple macros.

The following code does not work because the first macro (macroex1) calls the second macro (macroex2), and the second macro calls the first macro again.

```
DEFINE macroex1() RETURNS void {
..
..
Rel1 = macroex2();
 };
DEFINE macroex2() RETURNS void {
..
..
Rel2 = macroex1();
 };
```

No Macro Within Nested Block

A Foreach nested block cannot use macros. The following code is not valid:

```
Rel1=Foreach emp {
Rel = macroex();
Generate..
}
```

No Grunt Shell Commands

The Grunt shell commands are not allowed in a macro.

Invisible Relations

Relations defined within a macro are not visible to external Pig Latin code, and they cannot be referred to from external Pig Latin code.

The relations dnogrp and empcount are internal to macros.

```
DEFINE countagg(r,colname)  returns rerel{
dnogrp=group $r  by $colname;
empcount = foreach dnogrp generate group,COUNT($r.eno) as eno;
$rerel = filter empcount by group is not null;
}
```

Macro Examples

Here are some macro examples.

Macro Without Input Parameters Is Possible

A macro might not have any input parameters; however, such a macro is less dynamic. The following code removes a tuple when the *dno* column is null:

```
DEFINE filtenotnull() returns rerel{
$rerel = filter emp by dno is not null;
}
```

Macro Without Returning Anything Is Possible

A macro might not return anything. In such cases, you use void as its return type.

The following macro code stores the result into an HDFS directory so it does not need to return anything:

```
hdfs@cluster10-1:~> cat countagg.macro
DEFINE countagg(r,colname)  returns void{
dnogrp=group $r  by $colname;
empcount = foreach dnogrp generate group as group1,COUNT($r.eno) as eno;
store empcount into 'empcount';
};
```

You should not try to catch macro output because it is not returning anything. That is the reason you use a simple macro call as a last line without storing its output in a relation.

The following code uses a macro with a void return type:

```
hdfs@cluster10-1:~> cat importmacro.pig
import 'countagg.macro';
emp = load 'employee.csv' using PigStorage(',') as (eno:int,ename:chararray,
salary:int,dno:int);
countagg (emp,dno);
```

Summary

Macros in Pig Latin allow you to write more reusable and maintainable Pig Latin code. You have learned many things about macros in this chapter.

- You learned macro syntax and its usage in Pig Latin scripts.

- Macros can be two types: internal and external macros. An internal macro is part of Pig Latin script file, whereas an external macro is not.

- One macro internally referring to another macro is possible.

- You learned five important rules of writing macros.

- You also learned how to write parameterless macros and void return type macros.

CHAPTER 11

User-Defined Functions

In this chapter, you will learn how to write user-defined functions (UDFs) in Pig Latin.

I have discussed how Pig Latin provides two types of functions, as listed in Figure 11-1. There are many built-in functions, but those will not be sufficient for all requirements. Many times you need to write your own functions to fulfil your requirements. Such functions are called user-defined functions.

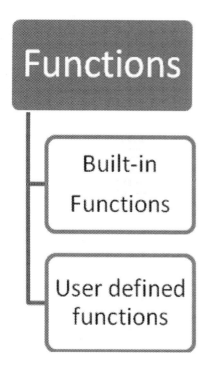

Figure 11-1. *The two types of function*

B. Vaddeman, *Beginning Apache Pig*, DOI 10.1007/978-1-4842-2337-6_11

User-Defined Functions

Apache Pig is a highly extensible platform and allows you to develop custom functionality using UDFs. Pig UDFs not only allow you to develop your own functions but also allow you to use a programming language you are comfortable with. It currently supports the programming languages Java, JavaScript, Python, Jython, Ruby, and Groovy, as listed in Figure 11-2.

Figure 11-2. *Pig's supported languages*

Java supports the complete functionality of UDFs, but support from JavaScript, Python, Jython, Ruby, and Groovy is still evolving. Cluster needs to have a runtime to use a programming language. As Java is a prerequisite for Hadoop, its runtime will be available by default. The JavaScript engine Rhino is available starting from Java 1.6 version. Most Unix systems will have Python installed by default, so the runtime will be available for Python. Other languages need to be installed explicitly for their runtime.

Java

You will learn how to write sample UDFs using Java and see how to use them in Pig Latin code.

The following are the steps for writing a UDF in Java.

Writing Java UDF

You need to extend the abstract class EvalFunc to create UDFs in Pig Latin. EvalFunc has an abstract method called exec that needs to be implemented to develop custom functionality.

You will write a UDF that replaces the null department number with the default department number 400.

The following code replaces the null value in the department column with 400:

```
package stringutils;
import java.io.IOException;
import org.apache.pig.EvalFunc;
import org.apache.pig.data.Tuple;

public class ReplaceNull extends EvalFunc<Integer> {
    public Integer exec(Tuple input) throws IOException {
        Integer dno = (Integer) input.get(0);
```

```
        try {
            if (dno == null)
            {
                return new Integer(400);
            }
        } catch (Exception e) {
            throw new IOException(
                    " exception thrown while processing department number ", e);
        }
        return dno;

    }
}
```

You will have tuple data inside the exec method, and you will use the get() method to get the first value from a tuple. If the first value is null, you will return 400. Here you have written this entire class in the stringutils package.

You can use the Eclipse IDE for writing a UDF. You need to have pig.jar in the build path to compile it successfully. After writing the UDF class, you need to create a JAR file.

Creating a JAR File

You can create a JAR file using a JAR program that comes with Java or you can use any IDE-like Eclipse. The Jar command needs to have the following syntax:

```
Jar cvf jarfilename file-names
```

Eclipse provides an export feature that can be used to export a list of Java files as a single JAR file. You need to right-click Project, click the Export option, and select Jar File under the Java category. After that, you will select .java files and enter the JAR file path where you want to store the JAR file. In the next screen, click Finish. In the last screen, it will ask you to select the main class; you do not need to select or enter the main class name as you do not have any main class in UDF.

You need to copy a JAR file to a Hadoop cluster if you have not generated on one of the cluster nodes. File transfer software such as WinScp or FileZilla can be used to transfer files from Windows to Unix.

Registering the Java UDF

You need to register the UDF JAR file to make it available during runtime. You use the register command to register the JAR file. The following register statement registers customutils.jar:

```
REGISTER customutils.jar;
```

The JAR needs to in the local file system. You can use both an absolute path or a relative path; the relative path will resolve from the current working directory in the local file system.

Using a Java UDF in Pig Latin Code

You can use user-defined functions in foreach generate like built-in functions. You will specify both the package name and the class name to use the UDF. The following code uses the replacenotnull UDF from the stringutils package.

```
replacenull = FOREACH emp GENERATE eno,ename,salary,stringutils.
replacenull(dno);
```

The complete code will look like this:

```
REGISTER strutils.jar;
Emp = load '/data/employee.csv' using PigStorage(',') as (eno:int,ename:char
array,salary:int,dno:int);
replacenull = FOREACH emp GENERATE eno,ename,salary,strutils.
replacenull(dno);
dump replacenull;
```

Rather than using a lengthy function name with a package structure, you can use a short alias name as a function. You use the DEFINE operator to define a short alias name as follows:

```
DEFINE replacenull strutils.replacenull();
```

This is simple eval function. You can write more advanced functions using Java. You will learn how to write more advanced functions in the next chapter.

JavaScript

You can write user-defined functions in JavaScript also. Pig Latin internally uses the JavaScript execution engine Rhino. Rhino is an open source framework that provides scripting capabilities to Java. You will learn how to write UDFs in JavaScript that change the employee name to uppercase.

Writing a JavaScript UDF

A JavaScript UDF contains two things. One is the output schema, and the other is the code of the JavaScript function. You specify the output schema of the JavaScript function, as shown here:

```
function-name.outputschema= "pig-latin-schema"
```

The following code specifies the output schema that contains one column of the varchar data type for a function called jsUpperCase:

```
jsUpperCase.outputSchema = "ename:chararray";
```

Now you will write a JavaScript function called jsUpperCase() that takes an employee name as input and returns the uppercased employee name. You can use the JavaScript function called toUpperCase() to do this.

```
function jsUpperCase(ename) {
    return ename.toUpperCase();
}
```

You will write these two (the schema and the JavaScript function) in a file. Also remember you can write multiple functions in a single file.

The following code contains a JavaScript UDF called jsUpperCase:

```
hdfs@cluster10-1:~> cat uppercasejs.js
jsUpperCase.outputSchema = " enameUC:chararray ";
function jsUpperCase(ename) {
     if(ename!==null)
    return ename.toUpperCase();
    else
    return ename;
}
```

Registering a JavaScript UDF with Pig

Once you have written a function in a file, you need to make it available during runtime. So, you need to register the JavaScript function using the register operator. You will use a JavaScript command or JsScriptEngine command as interpreter commands.

Both of the following register statements are valid:

```
register 'uppercasejs.js' using org.apache.pig.scripting.js.JsScriptEngine
as strfuncs;
```

```
register '/home/hdfs/uppercasejs.js' using JavaScript as strfuncs;
```

You can use both the relative path or the absolute path of the JavaScript function file in the register statement, and you will define an alias for this file using the AS keyword.

Using a JavaScript UDF in Pig Latin

You need to mention the alias name and function name to use it in Pig Latin code. The following code provides the employee name column to the JavaScript function jsUpperCase():

```
jsout = foreach emp generate eno,strfuncs.jsUpperCase(ename),salary,dno;
```

You can check the schema of the jsout relation that has gotten the new column name upperename, as mentioned in the JavaScript function file.

```
Describe jsout;
jsout: {eno: int,upperename: chararray,salary: int,dno: int}
```

The following code uses the JavaScript UDF in Pig Latin code:

```
hdfs@cluster10-1:~> cat uppercasejs.pig
register 'uppercasejs.js' using org.apache.pig.scripting.js.JsScriptEngine
as strfunc;
emp = load 'employee.csv' using PigStorage(',') as (eno:int,ename:chararray,
salary:int,dno:int);
jsout = foreach emp generate eno,strfunc.jsUCase(ename),salary,dno;
describe jsout;
```

Embedding Pig Latin Code

You can also embed Pig Latin code in JavaScript like you embed Pig Latin code in Java. You can use the JSPig class to do this. The compile() method of JSPig can be used to compile Pig Latin code.

The following code compiles Pig Latin code, and also Pig Latin code uses the JavaScript function:

```
var P = pig.compile(" emp = load 'employee.csv' as (eno:int,ename:chararray,
salary:int,dno:int);"+
    "jsout = foreach a generate eno,strfunc.jsUCase(ename),salary,dno"+
        "dump jsout;");
```

JSPig also provides a method called bind() to bind data to a variable. The bind() method returns an object of the BoundScript class that provides methods equal to the Pig Latin operators describe, explain, and run functionalities.

Other Languages

You will now learn how to write UDFs in other languages such as Jython, Python, Ruby, and Groovy.

Jython

You need to follow the same steps (write, register, and use) to use Jython code in the Pig Latin code as you do for writing Java and JavaScript UDFs. Writing a Jython UDF is the same as a JavaScript UDF; you need to write two things: one is the output schema, and other is the function code. The following code returns a number of characters in the employee name:

```
@outputSchema("word:chararray")
def  namelength(word):
  return len(word)
```

You can register Jython code using the `register` command along with the interpreter command. Jython or JythonScriptEngine can be used as the interpreter command.

The following two statements are valid `register` statements for Jython:

```
register 'namelen.py' using org.apache.pig.scripting.jython.
JythonScriptEngine as customeutils;
```

So is the following:

```
Register 'namelen.py' using jython as customeutils;
```

You can use the Jython function by using its name and defining its alias. The following code uses the Jython namelength function:

```
namelen = foreach emp generate customutils.namelength(ename);
```

The following is the complete Pig Latin code:

```
hdfs@cluster10-1:~> cat jythonfunc.pig
register 'namelen.py' using org.apache.pig.scripting.jython.
JythonScriptEngine as customutils;
emp = load 'employee.csv' using PigStorage(',') as (eno:int,ename:chararray,
salary:int,dno:int);
emp = foreach emp generate ename,customutils.namelength(ename);
dump emp;
```

Other Languages

Other language functions follow the same process as JavaScript and Jython. Functions contain two parts. One is the output schema, and the other is function code. The output schema is not mandatory. The default schema is taken if no output schema is specified.

You need to register functions using the `register` command and interpreter command. The interpreter command for Python could be `streaming_python` or `org.apache.pig.scripting.streaming.python.PythonScriptEngine`.

Table 11-1 contains the interpreter commands for all languages for quick reference.

Table 11-1. *Interpreter Commands for the Supported Languages*

Language	Interpreter commands	Examples
JavaScript	javascript,org.apache.pig.scripting.js.JsScriptEngine	register 'sample.js' using JavaScript as customutils or register 'sample.js' using org.apache.pig.scripting.js.JsScriptEngine as customutils
Jython	jython,org.apache.pig.scripting.jython.JythonScriptEngine	register 'sample.py' using Jython as customutils or register 'sample.py' using org.apache.pig.scripting.jython.JythonScriptEngine as customutils
Python	streaming_python,org.apache.pig.scripting.streaming.python.PythonScriptEngine	register 'sample.py' using Streaming_python as customutils or register 'sample.py' using org.apache.pig.scripting.streaming.python.PythonScriptEngine as customutils
Ruby	jruby,org.apache.pig.scripting.jruby.JrubyScriptEngine	register 'sample.rb' using jruby as customutils or register 'sample.rb' using org.apache.pig.scripting.jruby.JrubyScriptEngine as customutils
Groovy	groovy,org.apache.pig.scripting.groovy.GroovyScriptEngine	register 'sample.groovy' using Groovy as customutils or register 'sample.groovy' using org.apache.pig.scripting.groovy.GroovyScriptEngine as customutils

You can use other languages' UDFs in foreach generate statements like you use Jython and JavScript UDFs. You use the alias name set in the register statement and the function name defined in the code.

Other Libraries

Before writing your own UDFs, you can check out few more libraries such as PiggyBank and Apache DataFu. Both PiggyBank and Apache DataFu contain useful UDFs.

PiggyBank

PiggyBank is collection of UDFs written and shared by users. It comes as a JAR file when you download Apache Pig. PiggyBank UDFs contain evaluation UDFs and storage UDFs. Many useful storage UDFs such as CSVLoader, XMLLoader, and RegexLoader are available in PiggyBank. Also, many useful evaluation UDFs are available to process strings, dates, math, and XML data. You will use PiggyBank UDFs the same way you use normal UDFs.

The following code checks whether the input string is a numeric using the PiggyBank UDF IsNumeric:

```
REGISTER  piggybank.jar ;
Emp = load 'employee.csv' using PigStorage(',') as (eno: chararray,ename:cha
rarray,salary:int,dno:int);
Checknum = foreach emp generate org.apache.pig.piggybank.evaluation.
IsNumeric(eno);
Dump checknum;
```

PiggyBank UDFs may not have been tested well. They are available because they are shared by users.

Apache DataFu

Apache DataFu (https://datafu.incubator.apache.org/) provides a collection of libraries that are tested and suitable for large-scale data processing. It contains two libraries; one is for Pig, and the other is for Hourglass. Hourglass provides MapReduce libraries, and Apache DataFu Pig provides reusable UDFs that are well tested. DataFu's Pig library provides many UDFs for statistical analysis.

For example, you can generate the median for these listed numbers:

```
23
22
12
33
22
22
34
24
```

The following code uses DataFu's Median UDF to generate the median of the previous numbers:

```
register datafu-pig-incubating-1.3.0.jar
numrel = load 'numbers' using PigStorage() as (num:int);
grpall = group num all;
med = FOREACH grpall GENERATE Median(numrel.num);
DUMP med;
```

The previous code returns 22 as the median.

Summary

In this chapter, you learned to write UDFs in Pig Latin, including the following topics:

- How to develop UDFs using the Java language

- How to develop UDFs in the scripting language JavaScript and how to embed Pig Latin code in JavaScript code

- How to develop UDFs in other scripting languages such as Jython, Python, Ruby, and Groovy

- How to use PiggyBank UDFs

- How to use UDFs available in Apache DataFu

CHAPTER 12

Writing Eval Functions

In the previous chapter, you learned how to write user-defined functions. In this chapter, you will learn in detail how to write Eval functions using Java and how to access MapReduce features and Pig features inside Eval functions.

MapReduce and Pig Features

Let's start with MapReduce and Pig features.

Accessing the Distributed Cache

The MapReduce distributed cache is a useful feature to keep small files on all data nodes so that you can avoid costly operations like Reduce tasks. You can add files to the distributed cache using Pig. You can add both HDFS files and local files. Distributed cache files can be accessed within the exec method of Eval functions.

The following code adds the local file department.csv and the HDFS file loactions. csv to the distributed cache. It tries to access them within the exec method using their names.

```
public class DCAccessor extends EvalFunc<String> {

    @Override
    public String exec(Tuple input) throws IOException {
            Scanner scanlocalfile=new Scanner(new FileReader
            ("./department.csv "));

Scanner scanhdfsfile=new Scanner(new FileReader("./locations.csv "));

        return getdname("./department.csv", dno);
    }
@Override
    public List<String> getShipFiles() {
ArrayList<String> localfiles= new ArrayList<String>();
localfiles.add("/home/hdfs/department.csv");
```

© Balaswamy Vaddeman 2016
B. Vaddeman, *Beginning Apache Pig*, DOI 10.1007/978-1-4842-2337-6_12

```
        return localfiles;
    }
@Override
    public List<String> getCacheFiles() {
ArrayList<String> hdfsfiles= new ArrayList<String>();
localfiles.add("/home/hdfs/locations.csv");
        return localfiles;
    }
}
```

It is better to put small files in the distributed cache because it keeps files on every data node. It is useful for functionality like joins. You can get joins to access the distributed cache without using the join operator. For example, you can get the department name from the departments data set to get the department number in the employee data set easily from the distributed cache.

Accessing Counters

Counters display important statistics about the MapReduce job. Counters display many statistics. Some of them are the number of rows read, the number of rows written, the number of bytes read, and the number of bytes written. You can also access counters within a user-defined function. You can define new counters and also can retrieve built-in counters. You can use the PigStatusReporter class for both. This class provides a method called incrCounter that can be used to increase counters.

You will see how to write a new counter to count null values in the Department number column.

1. You need to create an enum with counter variables.

    The following enum called NullValues contains variables for counting null values in the department number and salary columns.

    ```
    package customutils;
    public enum NullValues {
    DNONULL,SALARYNULL
    }
    ```

2. You need to instantiate the PigStatusReporter class to call the incrCounter() method within the Eval function.

    The following code increases the value of DNONULL by 1 every time it gets a null value for the department number:

    ```
    public class ReplaceNull extends EvalFunc<Integer> {
        public Integer exec(Tuple input) throws IOException {
            PigStatusReporter psr=PigStatusReporter.getInstance();
            Integer dno = (Integer) input.get(0);
    ```

```
try {
    if (dno == null)
    {
        psr.incrCounter(NullValues.DNONULL, 1);
        return new Integer(400);
    }
} catch (Exception e) {
    throw new IOException(
            " exception thrown while processing
            department number ", e);
}
return dno;

        }
    }
```

After the job completion, you can check counters using the jobtracker URL, or you can simply run the following mapred command to see the MapReduce counters:

```
mapred job -status <job_id>
```

The command should display counters, and one of them will have the counter as shown here:

```
customutils.NullValues
            DNONULL=12
```

Reporting Progress

When you have user-defined functions that take a long time to complete, it is better to display the progress to the user. You can simply call the progress() method within the exec method of the Eval function. This will also avoid timeout errors in applications.

The following code calls the method progress() within the EVAL function:

```
public class ReplaceNull extends EvalFunc<Integer> {
    public Integer exec(Tuple input) throws IOException {
progress();
    }
}
```

Output Schema and Input Schema in UDF

You can define the output schema within a user-defined function. It is a good habit to define an output schema. The right schema will not only improve performance but also avoid unwanted failures.

You need to override the outputSchema() method of the EvalFunc class to define the output schema for the UDF.

The following code contains the outputSchema() method:

```
public Schema outputSchema(Schema input){
Schema udfSchema=new Schema();
..
..
Return udfSchema;

    }
```

This method has a schema of the input data; you can build an output schema using the same schema class if you want. As most of Pig Latin data types are the same as the Java data types, Pig can resolve the output data type if it is not specified in a user-defined function.

Table 12-1 contains the data type mapping between Pig and Java.

Table 12-1. *Data Type Mappings*

Pig Data type	Java Class
bytearray	DataByteArray
Chararray	String
int	Integer
long	Long
float	Float
double	Double
boolean	Boolean
datetime	DateTime
bigdecimal	BigDecimal
bigInteger	BigInteger
tuple	Tuple
bag	DataBag
map	Map

bag and tuple are simple Java interfaces. You cannot use them directly. You will use the TupleFactory and BagFactory classes to return the tuple and bag data types within UDF.

The following code creates a tuple with a capacity of ten fields and adds that tuple to a bag:

```
TupleFactory tfactory=TupleFactory.getInstance();
Tuple myTuple=tfactory.newTuple(10);

BagFactory bfactory= BagFactory.getInstance();
DataBag bag=bfactory.newDefaultBag();
bag.add(myTuple);
```

If the UDF does not provide any output schema, Pig will assume it is a single tuple with one field of bytearray. If the Pig Latin script does not honor this assumption, it will throw an error. Pig can get the data type of a variable using the Java Reflection API unless they are the bag and tuple data types.

The following code returns the correct schema even though the UDF replacenull did not override the outputSchema() method because ReplaceNull returns only Integer.

```
replacenull = FOREACH emp GENERATE strutils.replacenull(dno) as dno;
describe replacenull;

replacenull:{dno:int}
```

Examples of Output and Input Schemas

The following code defines a Tuple as the output schema:

```
public Schema outputSchema(Schema input) {
        Schema outSchema = new Schema(new Schema.FieldSchema(null,DataType.
        TUPLE));
        return outSchema;
```

The variable DataType.TUPLE represents the tuple data type in the schema. The DataType class contains all the variables that can be used for representing the schema.

You can also access the input schema within the UDF using the getInputSchema() method of the EvalFunc class.

The following code returns the first field from the input schema:

```
public Integer exec(Tuple input){
getInputSchema().getField(1);
}
```

Other EVAL Functions

You saw how to write simple EVAL functions in the previous chapter. Here you will learn how to write advanced EVAL functions such as aggregate, accumulator, and filter functions.

Algebraic

The EvalFunc class is used for writing both single-row functions and multirow functions. You will extend EvalFunc to write simple single-row functions, and if you want to write multirow functions such as an aggregate count and sum, you need to implement an extra interface called Algebraic. The Algebraic interface contains three methods to process initial data, intermediate data, and final data. Intermediate data processing is useful for reducing the load on the final stage. Algebraic functions take bags as input and produce scalar values as output.

The following code contains the Algebraic interface with its methods:

```
public interface Algebraic{
    public String getInitial();
    public String getIntermed();
    public String getFinal();
}
```

These three stages are same as the map, combine, and reduce stages in the MapReduce framework.

Table 12-2 matches the algebraic functions to the MapReduce functions.

Table 12-2. *Mapping the Functions*

Algebraic Function	Mapreduce Function
getInitial()	map
getIntermed()	combine
getFinal()	reduce

Initial Data Processing

Initial data processing is achieved by the class name returned by the method getInitial().

The following code returns the class name as CountInitial, which will actually have code for initial data processing:

```
Public string getInitial(){
Return CountInitial.class.getname();
}
```

The initial class has to extend EvalFunc and override the exec() method.

```
public class CountInitial extends org.apache.pig.EvalFunc<Tuple> {
@Override
    public Tuple exec(Tuple input) throws IOException {
.
.
}
}
```

getInitial is same as the Map task of the MapReduce framework. It is run only once per tuple. It takes a tuple as input and produces a tuple as output.

Intermediate Data Processing

Intermediate data processing is achieved by the class name returned by the method getIntermed().

The following code returns the class name as CountIntermed, which will actually have code for intermediate data processing:

```
Public string getIntermed(){
Return CountIntermed.class.getname();
}
```

The Intermed class has to extend EvalFunc and override the exec() method as shown here:

```
public class CountIntermed extends org.apache.pig.EvalFunc<Tuple> {
@Override
    public Tuple exec(Tuple input) throws IOException {
.
.
}
}
```

getIntermed() is same as the Combine task of the MapReduce framework. This will be called zero or many times. It takes a tuple as input and produces a tuple as an intermediate result.

Final Data Processing

The final data processing is achieved by the class name returned by the method getFinal().

The following code returns the class name as `CountFinal`, which will actually have code for intermediate data processing:

```
Public string getInitial(){
Return CountFinal.class.getname();
}
```

This `Final` class also extends the `EvalFunc` class and overrides the `exec()` method. `getFinal()` is same as the Reduce task of the MapReduce framework. This is the final stage, which will produce the final output.

```
public class CountFinal extends EvalFunc<Integer> {
@Override
    public Integer exec(Tuple input) throws IOException {
        return new Integer(sum);
    }
}
```

It will take a tuple as input but will produce a scalar value as output. The previous code is returning an integer as the final result. Now you will see how to write an algebraic function that counts null values.

Algebraic Function Example

An algebraic function should be computable in the three stages discussed. More importantly, it should be computable in the intermediate stage with partial data to reduce load on the final stage. For example, `Sum` is an algebraic function that can have the `SUM` function on the partial data in the intermediate stage and also the `SUM` function in the final stage.

The initial stage will have the default numbering most of the time like the `map` function in MapReduce.

Figure 12-1 shows the operations of the algebraic function `SUM` in three stages.

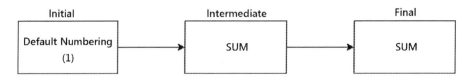

Figure 12-1. Algebraic function SUM

Sometimes you cannot have the same function in both the intermediate and final stages; in that case, you will plan a suitable function for the intermediate stage. For example, if you want to have a count of rows in a relation, you can use the count function in the intermediate stage, but you need to have `SUM` in the final stage, as shown in Figure 12-2.

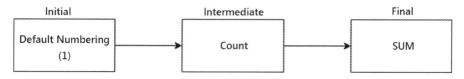

Figure 12-2. *Final stage*

Some functions such as ranking functions and statistical functions cannot be algebraic.

For example, say you want to count null values in a group. You will do the default numbering in the initial stage, which is 1. You will count the number of null rows in the intermediate stage, and you will sum all the intermediate values for the final sum. If all the rows after the initial stage have the default number 1, there is no difference in the count and sum in the intermediate stage.

Now you will see how to extend the EvalFunc class and implement the Algebraic interface to write an algebraic function that counts null values in a column.

Extend the EvalFunc Class

Like with the simple Eval function, you need to extend EvalFunc and override the exec() method. You will take the input tuple from the exec() method, will retrieve the first element as a data bag, and will iterate that data bag to get the final tuple and field value. If the field is null, you will increase the counter. You write this code as if there is no algebraic function implemented.

The following code contains the implementation for counting the null functionality:

```
@Override
    public Integer exec(Tuple input) throws IOException {
        int count = 0;

        DataBag db = (DataBag) input.get(0);
        if (db == null)
            return null;
        Iterator it = db.iterator();
        while (it.hasNext()) {
            Tuple dnotuple = (Tuple) it.next();
            for (int i = 0; i < dnotuple.size(); i++) {
                Object dno = dnotuple.get(i);
                if (dno == null) {

                    count++;
                }

            }

        }
        return new Integer(count);
    }
```

Implement the Algebraic Interface

You will now implement the Algebraic interface.

Writing Classes

You will write three classes, one for each stage. Every class will extend EvalFunc and implement the exec() method. You will write the initial class that will return 1 if null is found.

The following code is for the initial stage for counting nulls:

```
public class CountInitial extends org.apache.pig.EvalFunc<Tuple> {
    TupleFactory mTupleFactory=TupleFactory.getInstance();
    @Override
    public Tuple exec(Tuple input) throws IOException {
        DataBag bag = (DataBag)input.get(0);
        Iterator it = bag.iterator();
        if (it.hasNext()){
            Tuple t = (Tuple)it.next();
            if (t != null && t.get(0) == null)
                return mTupleFactory.newTuple(Integer.valueOf(1));
        }
        return mTupleFactory.newTuple(Integer.valueOf(0));
}
}
```

The intermediate class will return the number of rows that have one null. The following is the code for the intermediate stage to count the nulls:

```
public class CountIntermediate extends EvalFunc<Tuple> {
TupleFactory mTupleFactory=TupleFactory.getInstance();
int count = 0;
    @Override
    public Tuple exec(Tuple input) throws IOException {
            DataBag values = (DataBag)input.get(0);
        for (Iterator<Tuple> it = values.iterator(); it.hasNext();) {
                Tuple t = it.next();
                Integer i=(Integer)t.get(0);
            if(i.intValue()==1)
                count++;
        }
        Tuple medtuple=mTupleFactory.newTuple();
        medtuple.append(new Integer(count));
        return medtuple;
    }

}
```

The final class sums all the values in the input tuple to return the final result, which is the null count in this case.

The following code contains the final stage code for counting nulls:

```
public class CountFinal extends EvalFunc<Integer> {
    @Override
    public Integer exec(Tuple input) throws IOException {
        DataBag values = (DataBag)input.get(0);
        int sum = 0;
        for (Iterator<Tuple> it = values.iterator(); it.hasNext();) {
            Tuple t = it.next();
             if (t != null && t.size() > 0 && t.get(0) != null){
                Integer i=(Integer)t.get(0);
            sum += i.intValue();
            }
        }
        return new Integer(sum);
    }

}
```

Returning Classes

Now you need to implement methods of the Algebraic interface. As discussed earlier, the interface will have three methods that will return the class name of their respective stages.

The following code implements three methods. The getInitial() method returns the name of the initial class name that is getInitial, the getIntermed() method will return the intermediate class name CountIntermediate, and the getFinal() method returns the final class name that is getFinal.

The following code contains the implementation for the initial, intermediate, and final methods of the Algebraic interface:

```
public class AlgebraicCountNull extends EvalFunc<Integer> implements Algebraic {
public String getInitial() {
        return CountInitial.class.getName();
    }

    public String getIntermed() {
        return CountIntermediate.class.getName();
    }
public String getFinal() {
        return CountFinal.class.getName();
    }
```

```
@Override
public Integer exec(Tuple input) throws IOException {
..
..
}
```

Accumulator

All functions may not be algebraic. Statistical functions like median cannot be computed in multiple stages like in the Algebraic interface. These types of function will again increase the load on the Reduce task. The Accumulator interface will send data of the same key incrementally to reduce the load on the Reduce task. The Accumulator interface will also decrease memory consumption to boost application performance.

The Accumulator interface contains three methods, as follows:

```
public interface Accumulator <T> {
  public void accumulate(Tuple b) throws IOException;
 public T getValue();
  public void cleanup();
}
```

The following is the description for all three methods:

accumulate(Tuple b)

It is responsible for processing input tuple. It will be called one or more times.

getValue()

This will be called once after all the values of a key are passed to the accumulator.

cleanup()

It will be called after the getvalue() method to perform the cleanup operation.
The built-in aggregate function count implements the Accumulator interface.

Filter Functions

A filter function is a type of eval function that returns only a Boolean value. You need to extend the FilterFunc class and implement the exec() method to write your own filter function. The filter functions can be used in the Filter statement and also in the foreach generate statement.

The following code returns true if the column contains NULL or false otherwise:

```
import java.io.IOException;
import org.apache.pig.FilterFunc;
import org.apache.pig.data.Tuple;

public class IsNull extends FilterFunc {

    @Override
    public Boolean exec(Tuple input) throws IOException {
            Object obj=input.get(0);
        return (obj==null);
    }

}
```

You can use the IsNull function on the department number column to retrieve the NULL department number tuple, as shown here:

```
Emp = load 'employee.csv' using PigStorage() as (eo,ename,salary,dno);
Nulldno = filter emp by IsNull(dno);
Dump nulldno;
```

The output will be as follows:

```
(300,Nitya,150000,)
```

Summary

In this chapter, you learned to access MapReduce features and also learned to write advanced UDFs. Some of important topics are listed here:

- How to access files from the distributed cache within EVAL functions

- How to create new counters and access existing counters within the EVAL function using the PigStatusReporter class

- How to report the progress of a job using the progress() method

- How to access the input schema and define the output schema within the EVAL function

- How to write the algebraic function by implementing the getInitial, get Intermediate, and getFinal methods of the Algebraic interface

- The benefits of the Accumulator functions

- How to write a filter function by extending the FilterFunc class

CHAPTER 13

■ ■ ■

Writing Load and Store Functions

You have many load/store functions such as PigStorage, HBaseStorage, and TextLoader available in Pig, and many functions are available in PiggyBank. However, you may get other requirements to write your own load/store functions. You will learn how to write load and store functions in this chapter.

Writing a Load Function

You need to extend the abstract class LoadFunc and implement its abstract methods to write a simple loader function. The LoadFunc class can be used for loading data from many systems including Hadoop. The basic methods to be implemented are getNext(), getInputFormat(), prepareToRead(), and setLocation(), as listeds in Table 13-1, and the LoadFunc class comes with many methods with default implementations that you can even override if you want.

Table 13-1. Basic Methods to Implement

Abstract Methods
public Tuple getNext() throws IOException
public InputFormat getInputFormat()
public void prepareToRead(RecordReader reader, PigSplit split)
public void setLocation(String location, Job job)

© Balaswamy Vaddeman 2016
B. Vaddeman, *Beginning Apache Pig*, DOI 10.1007/978-1-4842-2337-6_13

- The getNext() method should return tuples one after another until all are finished. If you are reading data from Hadoop, it will use the Text class and the RecordReader class of Hadoop to read the next record and return it as a tuple.

- The prepareRead() method will set the split to be ready by Pig. It will be called before the getNext() method.

- The getInputFormat() method will return the input format class of MapReduce that is suitable for input files. MapReduce provides many input format classes; you can use one of them. If none of MapReduce input format classes is suitable, you need to write your own input format class using the MapReduce API.

Pig also provides input format classes that can be used as return classes of the getInputFormat() method. These Pig classes include all the files and directories recursively, unlike MapReduce.

PigTextInputFormat is a subclass of TextInputFormat and is used to read the text data type.

PigSequenceFileInputFormat is a subclass of SequenceFileInputFormat that is used to read the sequence file format data. PigFileInputFormat is a subclass of the FileInputFormat class and can be used to read any file. setLocation is used to set input paths, and you can even access the job details within this method. Internally the method relativeToAbsolutePath(String, Path)} will resolve relative paths if any are specified by the user. You will see how to load a double colon–delimited employee file using Pig Latin. You will override the four methods discussed.

The getnext() method will retrieve the next key available for processing using the method nextKeyValue(). If nextKeyValue() returns false, it means there are no keys to process, so the getNext() method returns null. The getCurrentValue() method returns a row that will be converted into the MapReduce Text type. Using text you will find the delimiter position. Using the delimiter position, you identify and read fields by data first into the array list, and you add that array list to a tuple. Every time it needs to return the fresh tuple, so you have to make it null in the end.

getInputFormat returns the PigTextInputFormat class, and the prepareRead() method will assign the RecordReader object to the local recordreader object using which you are reading data in the getnext() method.

The setLocation() method sets the input locations to the job using String inputPath.

Many MapReduce classes are involved in this program. MapReduce provides new classes in the MapReduce package and old classes in the mapred package. You need to refer to the new classes in the MapReduce package for import statements in the program.

This is a simple program with a double colon as the delimiter. It can be further developed to take any double characters as a delimiter.

The complete DoubleColonLoader class looks like this:

```
public class DoublecolonLoader extends LoadFunc {
    protected RecordReader reader = null;
    private String delimiter = "::";
    private ArrayList<Object> tupleList = null;
```

```
        private TupleFactory tupleFactory = TupleFactory.getInstance();
        @Override
        public Tuple getNext() throws IOException {
            try {
                boolean keyexists = reader.nextKeyValue();
                if (!keyexists) {
                    return null;
                }
                Text row = (Text) reader.getCurrentValue();
                int delimpos = row.find(delimiter);
                byte[] buf = row.getBytes();
                int len = row.getLength();
                int start = 0;

                for (int i = 0; i < len; i++) {
                    if (delimpos > 0) {
                        readField(buf, start, delimpos);
                        start = delimpos + 2;
                    }
                    delimpos = row.find(delimiter, start);
                }
                readField(buf, start, len);

                Tuple t = tupleFactory.newTupleNoCopy(tupleList);

                tupleList = null;
                return t;
            } catch (InterruptedException e) {
                int errCode = 6018;
                String errMsg = "Could not read input";
                throw new ExecException(errMsg, errCode,
                        PigException.REMOTE_ENVIRONMENT, e);
            }

        }
private void readField(byte[] buf, int start, int end) {
        if (tupleList == null) {
            tupleList = new ArrayList<Object>();
        }
        if (start == end) {
            tupleList.add(null);
        } else {
            tupleList.add(new DataByteArray(buf, start, end));
        }
    }
```

```
@Override
public InputFormat getInputFormat() {
    return new TextInputFormat();
}

@Override
public void prepareToRead(RecordReader rreader, PigSplit split) {
    reader = rreader;
}

@Override
public void setLocation(String inputpath, Job job) throws IOException {
    FileInputFormat.setInputPaths(job, inputpath);
}

}
```

You need to create a JAR file for the previous program and register it before using it, like any other user-defined function. After registering, you can use the class name as the function name along with the full package name.

The following code uses the DoubleColonLoader function in the Pig Latin script:

```
Register customutils.jar;
Emp = load 'employee' using customutils.DoubleColonLoader() as (eno:int,enam
e:chararray,salary:int,dno:int)
Enoename =foreach emp generate eno,ename;
Dump enoename;
```

Loading Metadata

You might want to load metadata about data if it is already available. You need to use the LoadMetadata interface to achieve this by implementing its four methods, listed in Table 13-2.

Table 13-2. *The LoadMetadata Methods to Implement*

Methods
public String[] getPartitionKeys(String arg0, Job arg1) throws IOException;
public ResourceSchema getSchema(String arg0, Job arg1) throws IOException ;
public ResourceStatistics getStatistics(String arg0, Job arg1);
public void setPartitionFilter(Expression arg0) throws IOException;

The getStatistics() method returns statistics about data to be loaded if any are available; otherwise, it returns null. The statistics feature for data is a work in progress. For now, you can simply use return null, as shown here:

```
@Override
    public ResourceStatistics getStatistics(String loc, Job job)
            throws IOException {
        return null;
    }
```

The getPartitionKeys() method returns partition keys available in the data; if no partition keys are available, it returns null. This is useful to read data from Hive tables. HCatLoader of Hive returns partition keys of the mentioned table if it is a partitioned table using this feature.

```
@Override
    public String[] getPartitionKeys(String loc, Job job)
    throws IOException {
        return null;
    }
```

You might have many partitions, but you may not be interested in all of them; in that case, you can filter the partitions you want using the setPartitionFilter() method. This method is a continuation of the getpartitionKeys() method, and it assumes partition keys returned by the getpartitionKeys() method are actually available. You will write an empty method if you do not have any partition filters like the following:

```
@Override
  public void setPartitionFilter(Expression pFilter) throws IOException {
}
```

You may not want to define a schema while loading data if the schema is already available in the input directory. You can simply specify -schema in the load statement so that it can automatically retrieve and define the schema for your relation. This functionality is achieved by implementing the getSchema() method.

The following Java code implements the getSchema() method of the LoadMetaData interface and returns the employee data set schema without defining it:

```
public ResourceSchema getSchema(String arg0, Job arg1) throws IOException {
        //define field and data type
        FieldSchema eno=new FieldSchema("eno", DataType.INTEGER);
        FieldSchema ename=new FieldSchema("ename", DataType.CHARARRAY);
        FieldSchema salary=new FieldSchema("salary", DataType.INTEGER);
        FieldSchema dno=new FieldSchema("dno", DataType.INTEGER);
        //add fields to list
        List<FieldSchema> fieldList=new ArrayList<Schema.FieldSchema>();
        fieldList.add(eno);
        fieldList.add(ename);
        fieldList.add(salary);
```

```
        fieldList.add(dno);
        //create a schema with list
        Schema empSchema=new Schema(fieldList);
        //return resource schema
        return new ResourceSchema(empSchema);
}
```

With the previous getSchema() method, the describe command will return the schema even though the schema is not defined by you.

The following code displays the schema of the relation emp even though the schema is not defined:

```
register customutils.jar;
emp = load 'employeedcolon.csv' using customutils.DoublecolonLoader() ;
describe emp;

emp: {eno: int,ename: chararray,salary: int,dno: int}
```

Improving Loader Performance

Rather than loading all the fields, it is a good idea to load only the required fields. This will improve the performance of the loader a lot. You can push down some fields to the loader using the LoadPushDown interface so that the loader loads those only fields. You need to implement two methods of the LoadPushDown interface, as noted in Table 13-3.

Table 13-3. The LoadPushDown Methods to Implement

Methods
public List<OperatorSet> getFeatures();
public RequiredFieldResponse pushProjection(RequiredFieldList arg0);

The getFeatures() method returns a set of operations that can be pushed down. The pushProjection() method informs the required field names to the loader.

Converting from bytearray

When you are writing the loader function, it is important not to lose data while reading it. Data will be available in byte format; you need to convert it to the appropriate data type. You can use the LoadCaster interface to convert the default data format (bytearray) to other Pig Latin data types.

For example, the following code converts bytearray to the Pig Latin chararray data type:

```
@Override
    public String bytesToCharArray(byte[] bytes) throws IOException {
        return new String(bytes);
    }
```

Table 13-4 contains all the methods to be implemented. You may not be required to implement all the methods. You may provide a skeleton implementation wherever required.

Table 13-4. *The Methods to Overload*

Methods
public Boolean bytesToBoolean(byte[] b) throws IOException;
public Long bytesToLong(byte[] b) throws IOException;
public Float bytesToFloat(byte[] b) throws IOException;
public Double bytesToDouble(byte[] b) throws IOException;
public DateTime bytesToDateTime(byte[] b) throws IOException;
public Integer bytesToInteger(byte[] b) throws IOException;
public String bytesToCharArray(byte[] b) throws IOException;
public Map<String, Object> bytesToMap(byte[] b, ResourceFieldSchema fieldSchema) throws IOException;
public Tuple bytesToTuple(byte[] b, ResourceFieldSchema fieldSchema) throws IOException;
public DataBag bytesToBag(byte[] b, ResourceFieldSchema fieldSchema) throws IOException;
public BigInteger bytesToBigInteger(byte[] b) throws IOException;
public BigDecimal bytesToBigDecimal(byte[] b) throws IOException;

For example, TextLoader implements only bytesToCharArray as it does not require the others.

Pushing Down the Predicate

Besides fields, you can also push down the operators and expressions you want. LoadPredicatePushdown allows you to perform advanced pushdown operations. You need to implement the three methods given in Table 13-5 to achieve this.

Table 13-5. *LoadPredicatePushdown Methods to Implement*

Methods
public List<String> getPredicateFields (String arg0, Job arg1) throws IOException ;
public List<OpType> getSupportedExpressionTypes ();
public void setPushdownPredicate (Expression exp) ;

This feature is still a work in progress; currently only ORCStorage implements it. You can define a set of operations to be pushed down using the getSupportedExpressionTypes() method. You can choose operations types such as EQUALS, NOT EQUALS using the OpType class.

Table 13-6 contains all the supported operations.

Table 13-6. *The Supported Operations*

Operation types
OpType.OP_EQ
OpType.OP_NE
OpType.OP_GT
OpType.OP_GE
OpType.OP_LT
OpType.OP_LE
OpType.OP_IN
OpType.OP_BETWEEN
OpType.OP_NULL
OpType.OP_NOT
OpType.OP_AND
OpType.OP_OR

Now you will learn how to write a custom store function.

Writing a Store Function

You need to implement StoreFuncInterface or extend the StoreFunc abstract class to write your own storage function. There is not much difference between them.

You use StoreFuncInterface if your class is already extending another class because Java does not allow you to extend two or more classes. StoreFuncInterface is especially useful for writing a single function for both loading and storing like the PigStorage function. In such a case, you will extend the LoadFunc class and implement the StoreFuncInterface interface. In either case, you need to implement the four methods shown in Table 13-7.

Table 13-7. *The Four Methods to Implement*

Methods
public OutputFormat getOutputFormat() throws IOException;
public void prepareToWrite(RecordWriter arg0) throws IOException;
public void putNext(Tuple arg0) throws IOException ;
public void setStoreLocation(String arg0, Job arg1) throws IOException;

The setStoreLocation() method will set the output location as the user input given in the store command of a Pig Latin script. Relative paths will be resolved using the method relToAbsPathForStoreLocation().

```
public void setStoreLocation(String loc, Job job) throws IOException {
    TextOutputFormat.setOutputPath(job, new Path(loc));
    }
```

The getOutputFormat() method will return the output format class. You can use the MapReduce output format classes such as FileOutputFormt, TextOutputFormat, and SequenceFileOutputFormat or the Pig output format classes such as the PigTextOutputFormat class. You can also create your own output format class using the MapReduce API.

The following code returns TextOutputFormat:

```
public OutputFormat getOutputFormat() throws IOException {
return new TextOutputFormat<WritableComparable,Text> ();
    }
```

The prepareToWrite() method is the starting point for storing data and called before the putNext() method.

The following code is an example of the implementation of the prepareToWrite method:

```
public void prepareToWrite(RecordWriter writer) throws IOException {
    rWriter=writer;
    }
```

The putNext() method is responsible for writing data to the output directory. This method runs once per tuple and writes one field after another using a delimiter in between them. It writes data in bytes after converting it into the appropriate data type.

The following store function writes data using a double colon as the delimiter. This is a simple program to write normal data types; you need to enhance it to support advanced data types.

```
public class DoubleColonStorer extends StoreFunc {
    private RecordWriter rWriter = null;
    private String delim = "::";
    private static final int BUFFER_SIZE = 1024;
    private static final String UTF8 = "UTF-8";

    @Override
    public OutputFormat getOutputFormat() throws IOException {
        return new TextOutputFormat<WritableComparable, Text>();
    }

    @Override
    public void prepareToWrite(RecordWriter writer) throws IOException {
        rWriter = writer;
    }

    ByteArrayOutputStream out = new ByteArrayOutputStream(BUFFER_SIZE);

    @Override
    public void putNext(Tuple tuple) throws IOException {
        int tsize = tuple.size();
        for (int i = 0; i < tsize; i++) {
            Object field;
            try {
                field = tuple.get(i);
            } catch (ExecException ee) {
                throw ee;
            }

            putField(field);

            if (i != tsize - 1) {
                out.write(delim.getBytes());
            }
        }

        Text text = new Text(out.toByteArray());
        try {
            rWriter.write(null, text);
            out.reset();
        } catch (InterruptedException e) {
            throw new IOException(e);
        }

    }
    private void putField(Object field) throws IOException {
```

```
        switch (DataType.findType(field)) {
        case DataType.NULL:
            break;
        case DataType.BOOLEAN:
            out.write(((Boolean) field).toString().getBytes());
            break;

        case DataType.INTEGER:
            out.write(((Integer) field).toString().getBytes());
            break;

        case DataType.LONG:
            out.write(((Long) field).toString().getBytes());
            break;

        case DataType.FLOAT:
            out.write(((Float) field).toString().getBytes());
            break;

        case DataType.DOUBLE:
            out.write(((Double) field).toString().getBytes());
            break;

        case DataType.BYTEARRAY: {
            byte[] b = ((DataByteArray) field).get();
            out.write(b, 0, b.length);
            break;
        }

        case DataType.CHARARRAY:
            out.write(((String) field).getBytes(UIF8));
            break;

        default: {
            int errCode = 2108;
            String msg = "Unknown Data type: " + field;
            throw new ExecException(msg, errCode, PigException.BUG);
        }

        }
    }

    @Override
    public void setStoreLocation(String loc, Job job) throws IOException {
        TextOutputFormat.setOutputPath(job, new Path(loc));
    }

}
```

This program extends the StoreFunc class. You can try implementing StoreFuncInterface. Please check the source code of PigStorage for a better understanding of this concept.

Writing Metadata

You need to implement the StoreMetadata interface by providing the implementation for the methods in Table 13-8 in order to store metadata in the output directory.

Table 13-8. StoreMetadata Methods to Implement

Methods
public void storeSchema(ResourceSchema arg0, String arg1, Job arg2);
public void storeStatistics(ResourceStatistics arg0, String arg1, Job arg2);

The storeSchema() method is used for storing metadata into the output directory. This method will create two hidden files, pig_header and pig_schema, under the output directory. The pig_schema file will have a schema of the relation name used in the store statement. And the schema will be stored in JSON format.

The following code stores metadata to the output directory:

```
public void storeSchema(ResourceSchema rSchema, String loc, Job job)
        throws IOException {
    JsonMetadata jsonMetadata=new JsonMetadata();
    jsonMetadata.storeSchema(rSchema, loc, job);

}
```

Once the job is completed, you can check the metadata files using the hdfs dfs -ls command that will list hidden files also.

The command below contains schema files .pig_schema and .pig_header.

```
hdfs@cluter10:~> hdfs dfs -ls empout
Found 4 items
-rw-r--r--   3 hdfs hdfs         21 2016-07-19 02:26 empout/.pig_header
-rw-r--r--   3 hdfs hdfs        426 2016-07-19 02:26 empout/.pig_schema
-rw-r--r--   3 hdfs hdfs          0 2016-07-19 02:26 empout/_SUCCESS
-rw-r--r--   3 hdfs hdfs         72 2016-07-19 02:26 empout/part-m-00000
```

If the script is run in local mode, you need to use the ls -a command that will list hidden files.

You can check the schema in the pig_schema file that will be in JSON format, as shown here:

```
hdfs@cluster10:~> hdfs dfs -cat empout/.pig_schema
{"fields":[{"name":"eno","type":10,"description":"autogenerated from Pig
Field Schema","schema":null},{"name":"ename","type":55,"description":"aut
ogenerated from Pig Field Schema","schema":null},{"name":"salary","type":10,
"description":"autogenerated from Pig Field Schema","schema":null},{"name":"dno",
"type":10,"description":"autogenerated from Pig Field Schema","schema":null}],
"version":0,"sortKeys":[],"sortKeyOrders":[]}
```

The storeStatistics() method is used to store statistics about data in the output directory. As discussed, loading/storing statistics about data is a work in progress; you can ignore this for now and just provide a skeleton implementation, as shown here:

```
public void storeStatistics(ResourceStatistics rStats, String loc, Job job)
            throws IOException {
}
```

Distributed Cache

The StoreResources class provides two methods to put files into the distributed cache (see Table 13-9). The getShipFiles() method puts local files into the distributed cache, and the getCacheFiles() method puts HDFS files into the distributed cache.

Table 13-9. *StoreResources Methods to Implement*

Methods
public List<String> getShipFiles ();
public List<String> getCacheFiles ();

The following Java code puts the departments.csv file of HDFS into the distributed cache:

```
public List<String> getCacheFiles(){
        String dfsdno="/user/hdfs/dfs-departments.csv";
        List<String> smallfiles=new ArrayList<String>();
        smallfiles.add(dfsdno);
        return smallfiles;

    }
```

The following Java code puts the departments.csv file of the local file system into the distributed cache:

```
public List<String> getShipFiles(){
    String localdno="/home/hdfs/local-departments.csv";
    List<String> smallfiles=new ArrayList<String>();
    smallfiles.add(localdno);
    return smallfiles;
}
```

As discussed earlier, you can avoid Reduce tasks by making use of the distributed cache. Most importantly, you will store only small files in the distributed cache.

Handling Bad Records

You can handle bad records or processing errors in a better way using the ErrorHandler and ErrorHandling interfaces. The ErrorHandler interface provides two methods, as listed in Table 13-10. They are onError and onSuccess. You will implement them to decide what to do when an error or a success occurs for a tuple.

Table 13-10. *ErrorHandler Methods to Implement*

ErrorHandler.java methods
public void onSuccess(String uniqueSignature);
public void onError(String uniqueSignature, Exception e, Tuple inputTuple);

The following two methods increase the counters when an error or a success happens so that you can check how many tuples failed and how many tuples succeeded:

```
public class StatusCounter implements ErrorHandler {
PigStatusReporter pigStatusReporter=PigStatusReporter.getInstance();
    public void onError(String arg0, Exception arg1, Tuple arg2) {
        pigStatusReporter.incrCounter("EmployeeRecords", "Failed", 1);
        }

    public void onSuccess(String arg0) {
        pigStatusReporter.incrCounter("EmployeeRecords", "succeeded", 1);
}
}
```

You need to implement the ErrorHandling interface in your UDF that returns the appropriate ErrorHandler class. Error-handling functionality will be performed without failing processing and when something goes wrong.

Accessing the Configuration

You can access the Hadoop configuration properties using the UDFContext class.

The following code gets a configuration object from UDFContext and gets a number of Map tasks launched for running the job using the property mapred.map.tasks:

```
UDFContext ctx= UDFContext.getUDFContext();
Configuration conf=ctx.getJobConf();
String maptasks=conf.get("mapred.map.tasks");
```

The configuration class provides several methods to handle the underlying properties. You can even modify or set a new value for a property.

The following code sets the number of Reduce tasks to ten:

```
conf.set("mapred.reduce.tasks","10");
```

Monitoring the UDF Runtime

You can monitor the UDF runtime so that you can time out long-running UDFs after a certain amount of time. Pig provides the annotation @MonitoredUDF that can be used to mark a UDF to monitor it.

```
@MonitoredUDF
public class SampleEval extends EvalFunc<Long> {
```

By default it will wait for ten seconds and will time out if the UDF runtime exceeds it. You can even provide an amount of time that a UDF can run before a timeout, as shown here:

```
@MonitoredUDF(timeUnit = TimeUnit.MILLISECONDS, duration = 100,
longDefault = 0)
public class SampleEval extends EvalFunc<Long> {
```

If a UDF times out, it will return the default value as 10. longDefault specifies the default value if the UDF return type is Long. Similarly, you can specify intDefault, DoubleDefault, stringDefault, and so on.

This feature is useful both for avoiding UDF going into an infinite loop and for improving UDF performance.

Summary

In this chapter, you learned many things about custom load/store functions, such as the following:

- How to write a custom load function to read double colon–delimited data

- How to load the metadata of a data set using the `LoadMetaData` interface

- How to load only specific fields using `LoadPushDown` to improve loader performance

- How to convert from one data type to another while loading data

- How to write a custom store function to write double colon–delimited data to an output directory

- How to write metadata using the `StoreMetadata` interface

- How to store small local files and HDFS files into a distributed cache

- How to do error handling

- How to set and get configuration properties

CHAPTER 14

Troubleshooting

Many times you might get stuck both while developing applications and while running applications. So, it is important to know how to troubleshoot Pig scripts. Pig provides features and operators for troubleshooting. You will learn about some of them in this chapter.

Illustrate

The Illustrate operator tells you how a Pig Latin script is transforming data. It generates sample data from the data set and displays how that data is transformed from one relation name to another. You can use the Illustrate operator on the relation name and also on the Pig Latin script. Illustrate will not work if the schema is not defined. It is not well maintained by the community, and many times it might not work. So, you should depend on describe when it does not work.

Here's the syntax:

```
Illustrate relname|scriptfile
```

Illustrate will also include the schema at every stage. See the following example:

```
grunt>emp = load 'employee.csv' using org.apache.pig.piggybank.storage.
CSVExcelStorage(',') as (eno:int,ename:chararray,sal:int);
grunt>grpall = group emp all;
grunt>cnt = foreach grpall generate COUNT_STAR(emp.eno);
grunt>illustrate cnt;
```

emp	eno:int	ename:chararray	sal:int	
	300	Nitya	1500001	
	200	Nirupam	2000002	

© Balaswamy Vaddeman 2016
B. Vaddeman, *Beginning Apache Pig*, DOI 10.1007/978-1-4842-2337-6_14

```
--------------------------------------------------------------------------
| grpall  | group:chararray  |emp:bag{:tuple(eno:int,ename:chararray,
sal:int)} |
--------------------------------------------------------------------------
|          | all               |{(300, Nitya, 1500001), (200, Nirupam,
2000002)}|
--------------------------------------------------------------------------

----------------------------
| cnt     | :long   |
----------------------------
|          | 2       |
----------------------------
```

describe

You might face schema-related issues while running Pig Latin scripts. describe is useful to correct them. As discussed earlier chapters, some operators like group, cogroup, and join will have their own schema. describe will help you understand their schema so that you continue writing appropriate code in the next lines of script.

Dump

You might face issues such as the Pig Latin script is not generating data. In such cases, you can use the DUMP operator to check which line is having a problem. You can also use it to check whether the script is generating the correct data or not.

Other operators such as LIMIT and SAMPLE can also be used for checking data in a relation.

Explain

The Explain operator displays a set of operations that explain how the Pig Latin script is converted and executed by Pig. This execution plan can be generated for one relation or the entire script.

Here's the syntax:

```
EXPLAIN [-script /path/to/pigscript] [-out /path/to/file] [-brief] [-dot]
[-xml] relname;
```

```
-script :
           Used to specify pig latin script path
-out :
           Used to specify file path where plan needs to be generated
```

```
-brief :
            Used to generate plan briefly

-dot :
            Used to generate DOT file to view plan in graph
-xml :
            Used to generate plan in XML format.
Relname :
            Relation name to be used for generating execution plan.
```

You now will see some examples for the explain operator.
The following code uses the explain operator in a relation from the Grunt shell:

```
Grunt>explain empcount;
```

The following code uses the explain operator in a Pig Latin script from the Grunt shell:

```
Grunt>explain -script /path/to/empcount.pig
```

The following code uses the explain operator in a Pig Latin script from a Unix prompt:

```
hdfs@cluster10:~> pig -x tez -e "expain -script /path/to/empcount.pig"
```

Plan Types

Three plans will be generated by the explain operator, as shown in Figure 14-1.

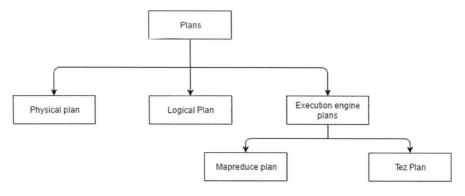

Figure 14-1. *The three plans*

The execution engine plan can be a MapReduce plan or a Tez plan depending on the execution engine you are using.

189

Logical Plan

A logical plan maps Pig Latin scripts to the internal logical operators of Pig. It will also have a schema at every stage. Not much optimization is performed in a logical plan. This is a simple mapping from a line of Pig Latin script to an internal logic operator.

Dotted lines represent the data flow, and the plan needs to be read from the bottom to the top. The bottom statement is a load statement, and the top statement is a store statement. Between the bottom and top statements, there are relation names, their associated logical operators, and the schema.

Figure 14-2 shows a sample logical plan.

```
cnt: (Name: LOStore Schema: countemp#67:long)
|---cnt: (Name: LOForEach Schema: countemp#67:long)
      (Name: LOGenerate[false] Schema: countemp#67:long)ColumnPrune:OutputUids=[67]ColumnPrune:InputUids=[63]
          (Name: UserFunc(org.apache.pig.builtin.COUNT_STAR) Type: long Uid: 67)
          |---(Name: Dereference Type: bag Uid: 66 Column:[0])
              |---emp:(Name: Project Type: bag Uid: 63 Input: 0 Column: (*))
          |---emp: (Name: LOInnerLoad[1] Schema: eno#50:int,ename#51:chararray,sal#52:int)
      |---grpall: (Name: LOCogroup Schema: group#62:chararray,emp#63:bag{#71:tuple(eno#50:int,ename#51:chararray,sal#52:int)})
          (Name: Constant Type: chararray Uid: 62)
          |---emp: (Name: LOForEach Schema: eno#50:int,ename#51:chararray,sal#52:int)
              (Name: LOGenerate[false,false,false] Schema: eno#50:int,ename#51:chararray,sal#52:int)ColumnPrune:OutputUids=[50, 51,
52]ColumnPrune:InputUids=[50, 51, 52]
                  (Name: Cast Type: int Uid: 50)
                  |---eno:(Name: Project Type: bytearray Uid: 50 Input: 0 Column: (*))
                  (Name: Cast Type: chararray Uid: 51)
                  |---ename:(Name: Project Type: bytearray Uid: 51 Input: 1 Column: (*))
                  (Name: Cast Type: int Uid: 52)
                  |---sal:(Name: Project Type: bytearray Uid: 52 Input: 2 Column: (*))
              |---(Name: LOInnerLoad[0] Schema: eno#50:bytearray)
              |---(Name: LOInnerLoad[1] Schema: ename#51:bytearray)
              |---(Name: LOInnerLoad[2] Schema: sal#52:bytearray)
          |---emp: (Name: LOLoad Schema: eno#50:bytearray,ename#51:bytearray,sal#52:bytearray)RequiredFields:null
*----------------------------------------------------------
```

Figure 14-2. *A sample logical plan*

Physical Plan

A physical plan will be generated after a logical plan is generated. A physical plan displays input paths and output paths. The structure of the physical plan will be the same as the logical plan; they both read from the bottom to the top, and the plan will be with respect to relation names.

A physical plan contains extra statements such as the local rearrange, the global rearrange, and the package. The local rearrange does the data preparation based on a key. The global rearrange is used for partitioning and shuffling. A package prepares the data required for a reducer.

Figure 14-3 shows a physical plan.

```
ent: Store(fakefile:org.apache.pig.builtin.PigStorage) - scope-20
:
:---ent: New For Each(false)(bag) - scope-19
   |   |
   |   POUserFunc(org.apache.pig.builtin.COUNT_STAR)(long) - scope-17
   |   |
   |   |---Project[bag][0] - scope-15
   |   :
   |   :---Project[bag][1] - scope-16
   |
   :---grpall: Package(Packages)[tuple][chararray] - scope-12
       |
       |---grpall: Global Rearrange[tuple] - scope-11
           :
           :---grpall: Local Rearrange[tuple]{(chararray)(false) - scope-13
           :   |
           :   Constant(all) - scope-14
           :
           :---emp: New For Each(false,false,false)(bag) - scope-10
               |   |
               |   Cast(int) - scope-2
               |   |
               |   |---Project[bytearray][0] - scope-1
               |   |
               |   Cast(chararray) - scope-5
               |   |
               |   |---Project[bytearray][1] - scope-4
               |   |
               |   Cast(int) - scope-8
               |   |
               |   |---Project[bytearray][2] - scope-7
               |
               |---emp: Load(hdfs://Cluster10/user/hdfs/employee.csv:org.apache.pig.piggybank.storage.CSVExcelStorage(',')) - scope-0
```

Figure 14-3. *A sample physical plan*

MapReduce Plan

Depending on the physical plan, Pig decides which operators should go in what stage of the MapReduce plan. A MapReduce plan has three more categories; they are map, combine, and reduce plans. The MapReduce plan is generated if you use MapReduce as the execution engine.

The map plan lists both map operations and also internal classes involved to prepare data required by the combiner and reducer stages. It also lists the loader function, file system details, and input path.

Figure 14-4 shows a map plan containing a foreach statement with cast operations and also COUNT_STAR#Initial, which does the data preparation required for the later stages. It contains the loader function CSVExcelStorage that is used to read data from the HDFS path /user/hdfs/employee.csv.

```
Map Plan
grpall: Local Rearrange[tuple]{chararray}(false) - scope-34
|
|   Project[chararray][0] - scope-36
|
|---cnt: New For Each(false,false)[bag] - scope-22
|   |   |
|   |   Project[chararray][0] - scope-23
|   |   |
|   |   POUserFunc(org.apache.pig.builtin.COUNT_STAR$Initial)[tuple] - scope-24
|   |   |
|   |   |---Project[bag][0] - scope-25
|   |       |
|   |       |---Project[bag][1] - scope-26
|   |
|   |---Pre Combiner Local Rearrange[tuple]{Unknown} - scope-37
|       |
|       |----emp: New For Each(false,false,false)[bag] - scope-10
|           |   |
|           |   Cast[int] - scope-2
|           |   |
|           |   |---Project[bytearray][0] - scope-1
|           |   |
|           |   Cast[chararray] - scope-5
|           |   |
|           |   |---Project[bytearray][1] - scope-4
|           |   |
|           |   Cast[int] - scope-8
|           |   |
|           |   |---Project[bytearray][2] - scope-7
|           |
|           |---emp: Load(hdfs://Cluster10/user/hdfs/employee.csv:org.apache.pig.piggybank.storage.CSVExcelStorage(',')) - scope-0-----------
```

Figure 14-4. *A sample map plan*

Tez Plan

A Tez plan is generated if you use Tez as the execution engine. The Tez plan will be the same as the MapReduce plan except that it lists the edges and vertices of the DAGs involved. Like MapReduce, the Tez plan lists three plans that display the initial class, intermediate class, and final classes of the Algebraic interface.

Figure 14-5 shows the first stage plan of the Tez plan.

```
#-----------------------------------------------------------
# TEZ DAG plan: pig-0_scope-0
#-----------------------------------------------------------
Tez vertex scope-21    ->    Tez vertex scope-22,
Tez vertex scope-22

Tez vertex scope-21
# Plan on vertex
grpall: Local Rearrange[tuple]{chararray}(false) - scope-35    ->    scope-22
|   }
|   Project[chararray][0] - scope-37
|---cnt: New For Each(false,false)[bag] - scope-23
|   |   |
|   |   Project[chararray][0] - scope-24
|   |   POUserFunc(org.apache.pig.builtin.COUNT_STAR$Initial)[tuple] - scope-25
|   |   |---Project[bag][0] - scope-26
|   |       |
|   |       |---Project[bag][1] - scope-27
|   |---Pre Combiner Local Rearrange[tuple]{Unknown} - scope-38]
|       |----emp: New For Each(false,false,false)[bag] - scope-10
|           |   Cast[int] - scope-2
|           |   |---Project[bytearray][0] - scope-1
|           |   Cast[chararray] - scope-5
|           |   |---Project[bytearray][1] - scope-4
|           |   Cast[int] - scope-8
|           |   |---Project[bytearray][2] - scope-7
|           |---emp: Load(hdfs://Cluster10/user/hdfs/employee.csv:org.apache.pig.piggybank.storage.CSVExcelStorage(',')) - scope-0
```

Figure 14-5. *First stage*

Modes

All explain plans can be generated in two modes, as shown in Figure 14-6. One is the text mode, and the other is the graph mode.

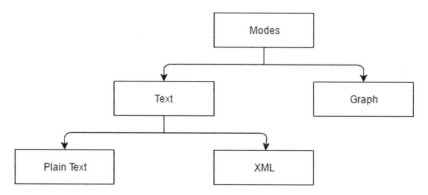

Figure 14-6. *The two plan modes*

Text mode can be further categorized into two modes: plain-text mode and XML mode. Plain-text mode is the default mode that generates all the explain plans in human-readable plain text, and by default the plan will be displayed on the console. The if -out option is used, and the plan will be saved in the file name mentioned.

You use the -xml option to generate the plan in XML format.

The following code generates all the plans in XML format for the relation empcount, and the partial XML is also shown:

```
grunt> explain -xml empcount;

<mapReducePlan>
  <mapReduceNode scope="25">
    <map>
      <POLocalRearrange scope="44">
        <alias>grpall</alias>
        <POProject scope="46"/>
        <POForEach scope="26">
          <alias>empcount</alias>
          <POProject scope="27"/>
          <POUserFunc scope="28">
            <POProject scope="29">
              <POProject scope="30"/>
            </POProject>
          </POUserFunc>
.
.
</map>
</mapreduce>
</mapreducePlan>
```

Graph mode generates code that can be used to view a plan in graph form. You can use any graph editor such as GraphViz to view a plan. You use the option -dot to generate the graph code.

The following code generates the explain plan in dot file format, and the partial logical plan code is also shown:

```
grunt> explain -dot empcount;
#------------------------------------------------
# New Logical Plan:
#------------------------------------------------
digraph plan {
compound=true;
node [shape=rect];
s602748972_in [label="", style=invis, height=0, width=0];
s602748972_out [label="", style=invis, height=0, width=0];
subgraph cluster_602748972 {
label="LOForEach"labelloc=b;
2010856706 [label="LOInnerLoad"];
2027549979 [label="LOInnerLoad"];
s210278405_in [label="", style=invis, height=0, width=0];
s210278405_out [label="", style=invis, height=0, width=0];
subgraph cluster_210278405 {
label="LOGenerate"labelloc=b;
1106933404 [label="Project0:(*)"];
1074868579 [label="Dereference"];
1906565212 [label="UserFunc"];
1106933404 -> 1074868579
1074868579 -> 1906565212
s210278405_in -> 1106933404 [style=invis];
969432090 [label="Project1:(*)"];
722764585 [label="Dereference"];
1657218305 [label="UserFunc"];
969432090 -> 722764585
722764585 -> 1657218305
s210278405_in -> 969432090 [style=invis];
};
```

You can save this dot code in a file and open it using a graph editor such as GraphViz. Figure 14-7 displays the partial logical plan in the GraphViz software.

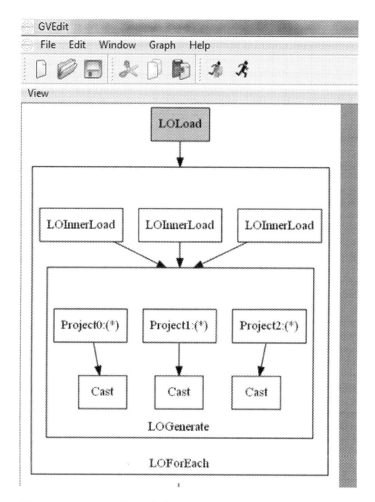

Figure 14-7. *Partial logical plan*

So far, you have learned about some operators for troubleshooting Pig Latin. Now you will learn some other concepts to troubleshoot Pig Latin code.

Unit Testing

Unit testing is useful during the development phase. Even a Pig Latin script can be unit tested. Though unit testing is time-consuming, it always improves the quality of reports. Unit testing can be done using the PigTest class in Pig. You will now learn how to write Java code for unit testing Pig Latin code.

Inside a Java class, you can use the @Test annotation to mark the Java method as a JUnit test method.

After that, you need to define some sample input data and the expected output data that can be used for testing the script. You can declare both the input and output as a Java string array.

The following code contains the input data:

```
String[] inputdata={"100,Bala,1000000,200",
           "101,Radha,1000000,300",
           "100,Nitya,1000000,100",
           "100,Pandu,1000000,"
                  };
```

The following code contains the expected output data:

```
String[] outputdata={"(4)"};
```

You need to instantiate the PigTest class. The PigTest class constructor can take Pig Latin code, Pig scripts, and cluster details.

The following code takes a Pig Latin script:

```
PigTest funtest = new PigTest("top_queries.pig", args);
```

PigTest provides many assert methods to check the expected output. The method assertOutput takes the input relation, output relation, input data, and output data. Pig runs scripts on the input data and generates the output data. If the generated output is the same as the expected output, it marks the unit test as a success and otherwise as a failure.

The following code contains the Pig Latin script:

```
hdfs@cluster10:~>cat /home/hdfs/empcount.pig
emp = load 'employee.csv' using PigStorage(',') as (eno:int,ename:chararray,
salary:int,dno:int);
grpall = group emp all;
empcount = foreach emp generate COUNT_STAR(emp.eno);
dump empcount;
```

The following is the complete Java code for unit testing the previously mentioned Pig Latin script:

```
package pigtest;
import java.io.IOException;
import org.apache.pig.pigunit.PigTest;
import org.apache.pig.tools.parameters.ParseException;
import org.junit.Test;
public class CountTester {
@Test
```

```
    public void testalgebraic(){
    try {

        String[] inputdata={"100,Bala,1000000,200",
                "101,Radha,1200000,300",
                "102,Nitya,1500000,100",
                "103,Pandu,5000000,"};
        PigTest pt= new PigTest("/home/hdfs/empcount.pig");

        String[] outputdata={"(4)"};
            pt.assertOutput("emp",inputdata,"empcount",outputdata);
} catch (IOException e) {
        e.printStackTrace();
    } catch (ParseException e) {
        e.printStackTrace();
    }
}

}
```

If the previous program generates 4 as its output, then the unit test is successful; otherwise, it is a failure. The previous program can run using Eclipse also. It requires many JAR files such as the Pigunit.jar, junit.jar, log4j, and hadoop JARs.

Error Types

Many Pig errors include error codes. It is good to know more about error codes. Pig throws four types of error codes. They are input, bug, user environment, and remote environment.

These are not only important to troubleshoot errors but are also useful for adding them in your applications for better exception handling.

An input error is thrown if there is an issue with the user input; most times these errors are thrown during syntax checking or if the script cannot be parsed.

For example, error code 1003, which is Unable to find operator for an alias, is thrown if Pig cannot resolve the operator.

Bugs are runtime errors. For example, error code 2043 is thrown if an unexpected exception occurs during data processing.

User environment errors are related to the current user environment. For example, error code 4010 is thrown if Pig cannot register the JAR file.

Most of the time, remote environment errors are related to other technologies such as Hadoop. For example, error code 6015 is thrown because of Hadoop errors.

Table 14-1 contains the error range for the previously discussed error types.

Table 14-1. *The Error Ranges*

Error Type	Error code range
Input	>=100 and <=1999
Bug	>=2000 and <===2999
User Environment	>=3000 and <=4999
Remote Environment	>=5000 and <= 6999

Counters

MapReduce counters are useful for troubleshooting both Hadoop issues and its abstractions like Pig. As discussed, Pig also provides support for counters; you can use the PigStatusReporter class. It is always a good habit to write your own counters to understand both data and data processing.

For example, to understand how many times methods of the Algebraic interface are called, you can simply add three counters to three methods, as shown here:

```
public void accumulate(Tuple input) throws IOException {
        pigStatusReporter.incrCounter("Accumulator", "accumulate",1);
.
.
.
}
public Integer getValue() {
pigStatusReporter.incrCounter("Accumulator", "getvalue",1);
.
.
.
}
public void cleanup() {
pigStatusReporter.incrCounter("Accumulator", "cleanup",1);
.
.
.
}
```

Accumulator is the group name of counters, and the counter names are accumulate, getvalue, and cleanup. You can use the MapReduce status command to see the counters. It displays both built-in counters and user-defined counters.

```
hdfs@cluster10:~>mapred job -status <job-id>
```

Here are some sample counters:

```
Map-Reduce Framework
                Map input records=69450395
                Map output records=69450395
                Map output bytes=763954345
```

```
                    Map output materialized bytes=230
                    Input split bytes=3680
                    Combine input records=69450395
                    Combine output records=10
                    Reduce input groups=1
                    Reduce shuffle bytes=230
                    Reduce input records=10
                    Reduce output records=1
                    Spilled Records=20
                    Shuffled Maps =10
                    Failed Shuffles=0
                    Merged Map outputs=10
                    GC time elapsed (ms)=139182
                    CPU time spent (ms)=633900
                    Physical memory (bytes) snapshot=16213774336
                    Virtual memory (bytes) snapshot=47089934336
                    Total committed heap usage (bytes)=15805710336
Accumulator
                    accumulate=1000
                    cleanup=1001
                    getvalue=1000
```

Summary

In this chapter, you learned to troubleshoot Pig Latin code using both Pig Latin operators and concepts, including the following:

- How to use the Illustrate operator to check the schema at every stage of the Pig Latin code

- How to use the describe operator to check the schema of a relation

- How to check the sample data of a relation using operators such as dump, sample, and limit

- How to use explain plans such as logical and physical, and the execution engine and modes (text and graph) of explain plans

- How to unit test Pig Latin code using the PigTest class

- That the four types of errors thrown by Pig Latin are user environment, input, remote, and environment

- How to write MapReduce counters using PigStatusReporter

CHAPTER 15

Data Formats

Storing and maintaining a huge amount data is one of the problems created by big data. In this chapter, you will learn how to store data efficiently using a few different data formats and compression algorithms.

Compression

The Hadoop ecosystem supports codec algorithms such as Gzip, LZO, Snappy, and Bzip2. A compression algorithm's primary responsibility is to reduce the size of data. You also need to check whether the algorithm supports splitting data; otherwise, the compression codec will not be suitable for parallel computing technologies like Hadoop.

Table 15-1 describes the codecs.

Table 15-1. Codecs

Algorithm	Splittable
Gzip	No
Bzip2	Yes
LZO	Yes after indexing
Snappy	No

- Gzip is not splittable, as it is available on Unix by default. You can use it if the data set's size is less than the block size.

- Bzip2 is splittable and has better performance than Gzip, but it is relatively slower than LZO and Snappy. It is also available by default on many Unix-based systems.

- You need to install the LZO program on most Unix systems, and it is not splittable by default but can become splittable after running an index program. It has good performance and can be used on data sets in the Hadoop ecosystem.

© Balaswamy Vaddeman 2016
B. Vaddeman, *Beginning Apache Pig*, DOI 10.1007/978-1-4842-2337-6_15

- Snappy is not splittable but has good performance. It is suitable for intermediate data processing in MapReduce.

Most of these algorithms can be auto-processed by Pig and Hadoop tools. They can auto-recognize codecs and process them.

It is also recommended that you apply compression for intermediate data for better performance of jobs because compressed data consumes less space and the file transfer is faster.

Sequence File

A sequence file is a binary file suitable for parallel computing.

It stores data in a key-value format. The first three characters, SEQ, say that it is a sequence file, and the next lines contain the records and header. The header includes the compression type used, the key and value classes, the metadata, and the sync marker. Figure 15-1 shows the structure of a sequence file format.

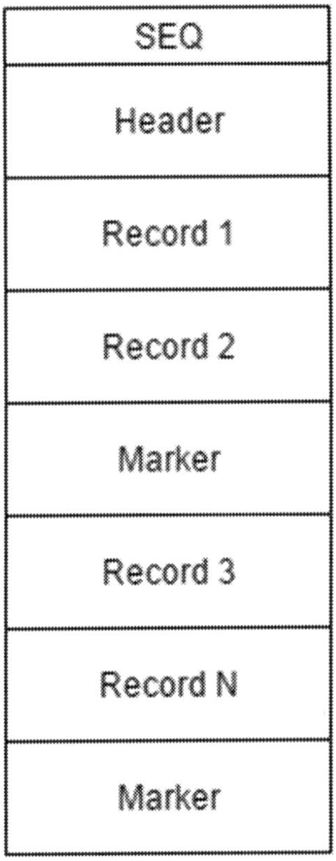

Figure 15-1. The sequence file format

The record contains the record length, the key length, the key, and the value. The value could be plain or compressed. The sequence file supports two types of compression; one is record-wise compression, and the other is block-wise compression. Record-wise compression is performed on a single record, whereas block-wise compression is performed on several records. Block-wise compression is preferred as it compresses the data size a lot.

It is a suitable data format for parallel computing because the marker allows HDFS clients to read data from multiple locations and often small files are combined to create sequence files to avoid performance issues. A normal cat program does not support it; you need to use the text command of the HDFS shell to see the content of the sequence file.

The following command displays the content of the sequence file:

```
hdfs dfs -text /path/to/seqfile
```

Pig Latin provides a multipurpose loader function called AllLoader that supports sequence files also.

Hadoop provides the classes SequenceFileInputFormt and SequenceFileOutputFormat that can be used to read and write sequence files. You can also use these classes to write your own load/store function in Pig Latin.

Hadoop provides one more data format called MapFile that is a sorted sequence file. The map file is sorted by key and also contains index data. Hadoop provides a class called MapFile to represent map files.

The map file will have better read performance because index data helps the HDFS client locate records easily.

Parquet

It is common to have a relation with a large number of columns, but a relation with a large number of columns requires a lot of time and multiple I/O calls to choose user-specified columns and process them. This problem is addressed by new data formats like Parquet.

Apache Parquet is a columnar storage-based data format that improves the performance of data processing and also decreases the cost of data storage.

For example, say you have the employee data set shown in Table 15-2.

Table 15-2. *An Employee Data Set*

eno	ename	salary	dno
200	Bala	1000000	101
201	Radha	1200000	102
202	Nitya	1500000	101

Row-based data formats will store data one row after another, as shown here:

| 200 | Bala | 1000000 | 101 | 201 | Radha | 1200000 | 102 | 202 | Nitya | 1500000 | 101 |

Parquet will store one column after another, as shown here:

| 200 | 201 | 202 | Bala | Radha | Nitya | 1000000 | 1200000 | 1500000 | 101 | 102 | 101 |

With the same type of data stored in one place, Parquet can provide type-based encoding. It can also perform better compression that will reduce greatly space consumption. Data is stored in row groups; if there is a condition in the query, the client can skip several row groups, which will result in less relation scanning. Parquet has good read performance because of a fewer number of I/O calls.

Initially it was developed by Twitter using the data format explained in the Dremel research paper; later Cloudera also contributed to Parquet. Now it is an open source project that belongs to the Apache Software Foundation. Parquet libraries are available in both Java and C++. Twitter says it has saved around 30 percent of storage, resulting in petabytes.

Parquet stores data in pages, and a page will contain header information, information about repetition levels and definition levels, and actual data. The data will be encoded as well as encrypted.

Parquet is not bound to any tool or technology and currently supports technologies such as Apache Hive, Cascading, Spark, Drill, Impala, and also Apache Pig. Many technologies such as Presto are in the process of supporting it. The Parquet format also works well on HDFS.

Parquet File Processing Using Apache Pig

Pig Latin provides the load function ParquetLoader to load Parquet files and the store function ParquetStorer to store output as Parquet files.

You need to register Parquet JARs to use both the ParquetLoader and ParquetStorer functions as these are not built-in functions yet.

The following code registers Parquet JARs and uses ParquetLoader:

```
register parquet-pig-1.2.2.jar ;
register parquet-encoding-1.2.5.jar;
register parquet-column-1.2.5.jar;
register parquet-common-1.2.5.jar;
register parquet-hadoop-1.2.5.jar;
register parquet-format-1.0.0.jar;
emp = load 'employee/' using parquet.pig.ParquetLoader as (eno:int,ename:cha
rarray,salary:int,dno:int);
```

The following code converts comma-separated data into the Parquet format:

```
emp = load 'employee.csv' using PigStorage(',') as (eno:int,ename:chararray,
salary:int,dno:int);
store emp into 'emp-parquet' using parquet.pig.ParquetStorer();
```

Files will have a .parquet extension, and the output folder will have a metadata file. The following command lists the output directory that contains the Parquet file:

```
hdfs@cluster10-1:~> hdfs dfs -ls  /home/hdfs/empparquet/
-rw-r--r-- 1 hdfs hadoop   0 Aug  1 02:35 _SUCCESS
-rw-r--r-- 1 hdfs hadoop 529 Aug  1 02:35 part-m-00000.parquet
-rw-r--r-- 1 hdfs hadoop 424 Aug  1 02:35 _metadata
```

The schema is considered as a message in Parquet. The field is defined using the field name, type, and repetition. The type could be a primitive data type like int, long, and string or a group that represents a nested data. The repetition can be one of three values: Optional, Required, and Repeated. Use Repeated to set the number of times it occurs.

The following code defines the employee data set schema:

```
Message employee{
Required int eno;
Optional String ename;
Optional group address{
    Optional int roadnumber;
    Optional string city;
    Repeated string city;
    Required int pin;
    }

}
```

Optional means zero or one occurrences. Required means only one occurrence. Repeated means zero or more occurrences.

Apache Parquet is mainly used for processing nested data. As Pig is also suitable for processing nested data, Parquet and Pig make a good combination for processing nested data.

Parquet is a relatively new project, and many features are on the road map.

ORC

Because of limitations in Record Columnar (RC) file format, new data format called Optimized Row Columnar (ORC) has been introduced in Apache Hive. ORC is a columnar file format that stores data in column after column instead of row after row. Initially it was written to improve the performance of Apache Hive, and now it is separate project in the Apache Software Foundation.

The ORC file format also mainly focuses on improving big data processing and also decreasing storage usage like Apache Parquet.

It is suitable for Internet-scale data processing and already proven in production at companies such as Facebook and Yahoo. Many technologies such as Apache Hive, Apache Pig, Presto, and Spark can support the ORC format.

As you learned in Chapter 5, Apache Pig supports the ORC format using the ORCStorage function that is used for both loading and storing ORC file formats.

The default compression used by ORCStorage is ZLib.

```
emp = LOAD 'employee.orc' USING OrcStorage();
```

Please check Appendix C for more options in the ORCStorage function.

Each ORC file is divided into stripes. And each stripe contains index information, actual data, and a footer. At the end of stripes, there will be a file footer and PostScript. The stripe works the same as an HDFC block, but its size will be bigger than the block. The stripe default size is 256 MB, but you can change it.

Figure 15-2 shows the structure of the ORC file format.

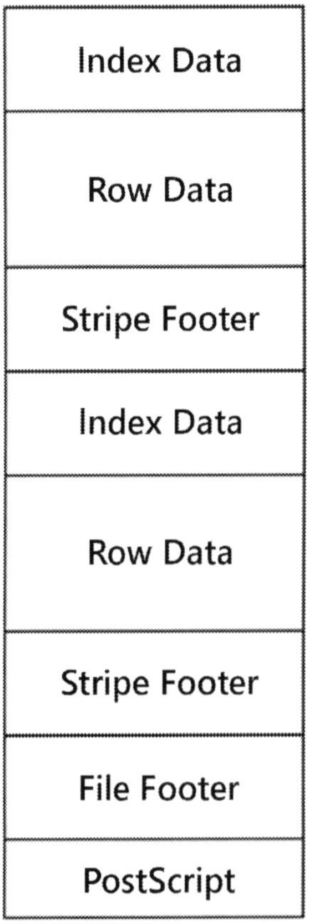

Figure 15-2. *The ORC file format*

Index

Indexes are maintained at three levels: the file, stripe, and row levels. File and stripe indexes maintain statistics about files and stripes, and these are stored in the file footer. The row index contains statistics about the row group and also the start position of the row group.

ACID

HDFS follows a "write-once, read-many-times" philosophy. If you have to update some data, it is not a single straight step. You need to first delete and then create a file with the new data. It is the same case for Apache Hive also.

The ORC file format takes Apache Hive to the next level by supporting updates and also the atomicity, consistency, isolation, and durability (ACID) properties. For example, streaming applications such as Flume, Kafka, and Storm can create transactions in Apache Hive using the ORC file format.

The ORC file supports the ACID properties using base and delta files. For all updates, delta files are created. If the delta files are large enough, they are rewritten to base directories.

Transaction support in Hive requires the following properties to be present:

```
Set hive.support.concurrency ='true'
Set hive.enforce.bucketing ='true'
Set hive.exec.dynamic.partition.mode = nonstrict
Set hive.txn.manager = org.apache.hadoop.hive.ql.lockmgr.DbTxnManager
Set hive.compactor.initiator.on =' true'
Set hive.compactor.worker.threads =120;
```

Predicate Pushdown

The best part of the ORC format is that it can push down predicates to stripes. The pushdown operations are supported by many tools such as Apache Hive, Cascading, and Apache Pig. For example, if you have any condition after a load statement in Pig, it can directly load data from the file that matches with the condition.

Data Types

ORC supports several data types such as integers, floats, strings, complex types, and binary types.

Figure 15-3 lists the data types of ORC.

Integer	Float	String	Date	Complex	Blob
boolean	Float	String	Timestamp	struct	Binary
tinyint	Double	char	Date	list	
smallint		varchar	·	map	
int				union	
bigint					

Figure 15-3. *The ORC data types*

Benefits

The ORC file format is a more advanced file format than the RC and Parquet file formats. These are some important benefits of ORC files:

- The best feature of ORC is its support for transactions.

- ORC has better read performance than Parquet as it stores statistics about data.

- It reduces the data size better than Parquet.

Summary

You learned many important things about data formats in this chapter, including the following:

- Codecs such as Gzip, Bzip2, LZO, and Snappy were covered.

- The sequence file format is suitable for parallel computing because it is splittable.

- A map file will have better read performance than a sequence file because it maintains index data.

- Apache Parquet is a columnar storage-based data format that improves the performance of big data processing and decreases the cost of big data storage.

- Apache ORC is a columnar data format that supports the ACID properties in Apache Hive.

- Apache ORC supports predicate pushdown to load only the required columns from the input data set.

- Apache ORC provides better performance and consumes less space than Apache Parquet.

CHAPTER 16

Optimization

In big data processing, performance is important so that people can make quicker decisions based on the available reports. You should not be happy with just having output; you should also check how much time you have taken for that output and should try decreasing the running time.

In this chapter, you will learn some optimization tips.

Advanced Joins

A join operation is a Reduce task, and it is one of the most costly operations in Pig Latin. So, you need to take extra care with joins to improve their performance.

Pig Latin provides some advanced join operations that will provide better performance.

Small Files

When some of your relations are small enough to fit into memory, Pig Latin will load them into memory and read from there to perform the join. These are called *replicated joins*. You will use the `replicated` keyword in the `join` operator to enable a replication join.

The following code contains replicated joins:

```
emp = load 'employee.csv' using PigStorage(',') as (eno:int,ename:chararray,
salary:int,dno:int);
dept = load 'department.csv' using PigStorage(',') as
(dno:int,dname:chararray);
empjn = join emp by dno,dept by dno using 'replicated';
```

Replicated joins are faster than normal joins because they do not require you to perform a Reduce task. The maximum size of a relation that can be put into memory is decided by the property `pig.join.replicated.max.bytes`. Its default value is 1 GB. If this property value is exceeded, the join will fail.

© Balaswamy Vaddeman 2016
B. Vaddeman, *Beginning Apache Pig*, DOI 10.1007/978-1-4842-2337-6_16

User-Defined Join Using the Distributed Cache

When you do join two data sets, MapReduce copies the output of the two Map tasks to another node where the join is performed in a Reduce task.

For example, if you want to join the employee and department data sets, the output of the employee Map task and the output of the department Map task are sent to the reduce node where it performs the join functionality, matching the data from both map outputs.

A replicated join puts one or some of the data sets into the distributed cache so that it can read the other data set from the same map node and it does not require any Reduce task. If you can avoid a Reduce task, you can improve performance.

You can write your own Eval function for the join functionality that avoids the Reduce task just like with a replicated join. In fact, you can have more control over the joins by writing your join function. You will not have any size constraint.

Here you will write an Eval function that performs a join on the employee relation and department.csv using the distributed cache.

The following is the sample employee data set:

```
Employee.csv
----------------------
300,Nitya,1500001,100
301,bala,120001,200
302,pandu,1300002,300
303,balu,1000000,400
304,Nirupam,100000,
```

The following is the sample department data set:

```
Department.csv
----------------------
100,it
200,support
300,hr
400,admin
```

The following code performs the join operation using the distributed cache:

```java
public class CacheJoin extends EvalFunc<String> {

    @Override
    public String exec(Tuple input) throws IOException {
        Integer dno = (Integer) input.get(0);

        if (dno != null)
            return getdname("./department.csv", dno);
        else
            return "OTHERS";
    }
```

```
    String getdname(String filename, int dno) throws FileNotFoundException {
        Scanner sc = new Scanner(new FileReader(filename));
        HashMap<Integer, String> map = new HashMap<Integer, String>();
        while (sc.hasNext()) {
            String[] values = sc.nextLine().split(",");
            map.put(new Integer(values[0]), values[1]);
        }
        String dname = map.get(new Integer(dno));
        if (dname != null) {
            dname = dname.toUpperCase();
        }
        return dname;

    }

    @Override
    public List<String> getCacheFiles() {
        ArrayList<String> dfsfiles = new ArrayList<String>();
        dfsfiles.add("/user/hdfs/department.csv");
        return dfsfiles;
    }

}
```

It also converts the department name into uppercase. And it returns OTHERS as the department name if no department number is in the employee relation.

- The getdname() method reads the department.csv file from the distributed cache and loads the department number and department name into the HashMap. It retrieves the department name for the department number and converts it into uppercase.

- The exec method returns the department name in uppercase if the department number is not null or returns OTHERS as the default department name.

The following code uses the CacheJoin function, which does not load department. csv using the load statement, and no join operator is used. Still, it achieves the join functionality.

```
Pig Latin Script
--------------------
register customutils.jar;
emp = load 'employee.csv' using PigStorage(',') as (eno:int,ename:chararray,
salary:int,dno:int);
dnodname =foreach emp generate ename,customutils.CacheJoin(dno);
dump dnodname;
```

OUTPUT
```
--------------------
(Nitya,IT)
(bala,SUPPORT)
(pandu,HR)
(balu,ADMIN)
(Nirupam,OTHERS)
```

Big Keys

Sometimes it happens that one key will get a big volume of data and others will get less data in the reduce stage. In such skewed data cases, the reduce node that is processing the key with the big volume of data will have slow performance; sometimes it might even fail the job.

Pig Latin provides an advanced join called a *skewed* join to address these skewed data cases.

The following code uses a skewed join:

```
emp = load 'employee.csv' using PigStorage(',') as (eno:int,ename:chararray,
salary:int,dno:int);
dept = load 'department.csv' using PigStorage(',') as (dno:int,dname:chararray);
empjn = join emp by dno,dept by dno using 'skewed';
```

- The skewed joins work only with two-way joins, and the skewed relation needs to be the first one.

- If more than two relations are involved in the join, it is better to convert them into two-way joins.

- pig.skewedjoin.reduce.memusage allows you to choose the amount of memory for the join operation. The default value is 0.5 GB. You can modify it and check the performance of the join.

- The primary goal of a skewed join is to avoid job failures.

Sorted Data

If the data is already sorted for both relations, Pig Latin provides an advanced join called a *sort-merge* join that will avoid the Reduce task when joining two data sets. This will improve the performance of the joins significantly because only the map phase is involved and no sort and shuffle phases are required.

This sort-merge algorithm takes the right-side relation data as the side data, builds an index on it, and performs a join with the left-side relation data by matching the left-side data to the index data. There are many prerequisites for the successful completion of the sort-merge join.

The following code uses a sort-merge join:

```
emp = load 'employee.csv' using PigStorage(',') as (eno:int,ename:chararray,
salary:int,dno:int);
dept = load 'department*.csv' using PigStorage(',') as (dno:int,dname:chararray);
empjn = join emp by dno,dept by dno using 'merge';
```

The following are the rules to be followed by a sort-merge join:

- Data coming to the join statement should be from the load or order by statement.

- Data needs to be sorted by common key in ascending order.

- If there are any statements between join and order by, they should not change the order of columns.

These four join strategies are useful for improving the performance of join operations.

Best Practices

Now you will learn some best practices to be followed while writing Pig Latin code.

Choose Your Required Fields Early

You need to choose your required fields as early as possible so that you are not processing unnecessary data that consumes I/O, CPU time, and sometimes disk space.

For example, it is a good idea to use a foreach statement immediately after a load statement, as follows, instead of directly using group operations:

```
emp = load 'employee.csv' using PigStorage(',') as (eno:int,ename:chararray,
salary:int,dno:int);
emp = foreach emp generate eno,dno;
grp = group emp by dno ;
grpcnt = foreach grp generate FLATTEN(emp.dno),COUNT(emp.eno);
```

Define the Appropriate Schema

If a schema is not defined, Pig Latin will assign a default schema and will perform type casting depending on the requirements.

For example, if you do not define a schema along with a load statement, Pig Latin will assign the default data type bytearray and will convert to the double data type if there are any arithmetic operations like in the following code:

```
Emp = load 'employee.csv' using PigStorage(',');
Newsal = foreach emp generate $2,$2*1.1;
```

If you define a schema, you can avoid such internal operations. It will also be helpful for query planning internally.

Unlike the earlier example, the following code defines the schema:

```
Emp = load 'employee.csv' using PigStorage(',') as (eno:int,ename:chararray,
salary:int,dno:int);
Newsal = foreach emp generate salary,salary*1.1;
```

Filter Data

It does not make sense to perform operations on unnecessary data. It will only increase the running time of the job.

For better performance, you should remove data as early as possible and as many times as possible.

The following code performs a filter after a group by operation:

```
emp = load 'employee.csv' using PigStorage(',') as (eno:int,ename:chararray,
salary:int,dno:int);
grp = group emp by dno ;
grpcnt = foreach grp generate FLATTEN(emp.dno),COUNT(emp.eno);
grpcnt = filter grpcnt by dno!=100;
```

You can move up the filter statement so that the group operation will process relatively less data.

The following code moves the filter up and improves the performance of the job:

```
emp = load 'employee.csv' using PigStorage(',') as (eno:int,ename:chararray,
salary:int,dno:int);
emp = filter emp by dno!=100;
grp = group emp by dno ;
grpcnt = foreach grp generate FLATTEN(emp.dno),COUNT(emp.eno);
```

Store Reusable Data

You will perform cleaning operations in most of your data-processing projects. Cleaning operations will remove tuples that are actually not valid. If you have 20 reports to be generated, there is no point in performing cleaning operations in every report unless they are specific to a report.

You need to store reusable data so that you can remove redundant operations. Of course, maintaining the staging directory will consume extra space on the cluster.

Use the Algebraic Interface

Instead of directly going to the Reduce task from the Map task, it is better to have a Combine task that will reduce the amount of data to be written to disk and also the load on internal operations such as sort and shuffle.

The `Algebraic` interface contains three stages: initial, intermediate, and final. These cover the Map, Combine, and Reduce tasks.

- The performance of the `Algebraic` job will be improved if you use the `Algebraic` function.

- You should also try to implement the `Algebraic` interface whenever you are writing user-defined functions.

Use the Accumulator Interface

Not all functions fit into the `Algebraic` interface. So, to reduce load on the Reduce task, you can implement the `Accumulator` interface.

As discussed earlier, `Accumulator` will send data to the Reduce task in parts rather than all at once.

You should consider using the `Accumulator` interface if you are writing an UDF. You can also use both the `Algebraic` and `Accumulator` interfaces at the same time.

Compress Intermediate Data

MapReduce writes intermediate data to disk and copies to the appropriate reduce node from disk. There will be huge I/O operations involved, depending on the amount of intermediate data.

It is a good idea to compress intermediate data for better performance. You can use LZ0, Gzip, and Snappy codecs to do this.

You should first set the compression setting `mapred.compress.map.output` to true; then you can specify the codec using `mapred.map.output.compression.codec`.

The following code enables compression for intermediate data:

```
set mapred.compress.map.output 'truc' ;
set mapred.map.output.compression.codec ' org.apache.hadoop.io.compress.LzoCodec';
```

Pig also supports temporary file compression, and you can use the following properties for better performance of a job:

```
set pig.tmpfilecompression 'true';
set pig.tmpfilecompression.codec 'Lzo';
```

Combine Small Inputs

Hadoop is most suitable for big files and less suitable for small files. Small files increase disk seek operations that will impact job performance.

You should combine small files into one so that a single map can process them. Pig Latin provides properties for such combine functionality.

The following code sets the maximum amount of memory as 120 MB. Small files will be combined up to 120 MB. The value is specified in bytes.

```
set pig.maxCombinedSplitSize '120000000'
```

Prefer a Two-Way Join over Multiway Joins

It is always a good idea to convert multirelation joins into two-way joins. This will help you understand which join is taking more time. Also, it is suitable for applying advanced joins like skewed joins. Two-way joins will also reduce the load on Reduce tasks, and you will have fewer job failures.

Better Execution Engine

Apache Tez addresses the problems of MapReduce and also provides better resource management such as container reuse. As it enables interactive data analysis, it is recommended that you use Tez as an execution engine. In future releases, you can even use Spark as the execution engine, which provides in-memory cluster computing.

The following code uses the Tez execution engine:

```
Pig -x tez -f /path/to/script.pig
```

Parallelism

The number of reducers is decided by the execution engine. Pig Latin also provides properties to change the number of reducers.

The following code sets the data size per reducer:

```
set pig.exec.reducers.bytes.per.reducer '1200000000'
```

The following code sets the number of reducers:

```
set pig.exec.reducers.max '50'
```

Pig Latin will decide on the number of reducers based on the previous property values and available data size for the reduce operation. It uses the following formula to decide on the number of reducers:

```
Number of reducers = MIN(pig.exec.reducers.max, input_data_size_in_bytes/
bytes_per_reducer)
```

If you have the maximum number of reducers set to 20, the input data size as 50 GB, and the amount of data per reducer as 1 GB, Pig Latin will launch 20 reducers.

```
Number of reducers = MIN (20,50 (GB)/1(GB))
Number of reducers  = 20
```

In a Pig Latin script, you can specify the number of reducers at the script level and also at the operator level.

default_parallel sets the number of reducers at the script level, as shown here:

```
set default_parallel 50;
```

Many operators such as CROSS, JOIN, GROUP, COGROUP, and ORDER allow you to set the number of reducers in Pig Latin. You can use the parallel keyword for them separately.

The following code sets the number of Reduce tasks as 10 for the order by operator:

```
Sortdno = order emp by dno parallel 10;
```

Job Statistics

It is a good habit to gather job statistics once a job finishes. You can gather information such as how much data is processed, what data format is used, what operators are used, and how much data is produced as output.

You can gather statistics using the counters generated by the job. The counters will provide useful information such as how much data is spilled and how much data is processed by the sort and shuffle tasks. When you make changes to a script, you can check how changes are impacting internal tasks.

You can even enable useful counters for a Pig Latin script to check the performance of a UDF.

- You can set the property pig.udf.profile to true so that Pig Latin will generate performance-related counters.

- The approx_invocations counter tells you how many times the UDF is called.

- The approx_microsecs counter will tell you the time taken by the UDF.

The following code enables performance-related counters:

```
set  pig.udf.profile true;
emp = load 'employee.csv' using PigStorage(',') as (eno:int,ename:chararray,
salary:int,dno:int);
emp = filter emp by dno!=100;
grp = group emp by dno ;
grpcnt = foreach grp generate FLATTEN(emp.dno),COUNT(emp.eno);
dump grpcnt;
```

The following are performance-related counters of the UDF called COUNT:

```
org.apache.pig.builtin.COUNT
                approx_invocations=100
                approx_microsecs=29900
```

Rules

Pig provides many rules to optimize user scripts for better performance. All of these rules are enabled by default. You can disable some rules if you want.

The following code disables `ConstantCalculator` and `Splitfilter`:

```
set pig.optimizer.rules.disabled ' ConstantCalculator,Splitfilter'
```

You can also use `optimizer_off` or the -t option to disable rules.

The following are examples to disable all rules:

```
pig -optimizer_off all -x local -f grpcnt.pig
pig -t all -x local -f grpcnt.pig
```

You can see a message on the console that says which rules are enabled and which rules are disabled after running the script.

The following shows the `RULES_ENABLED` and `RULES_DISABLED` lines:

```
2016-07-28 11:20:02,653 [main] INFO org.apache.pig.newplan.logical.
optimizer.LogicalPlanOptimizer - {RULES_ENABLED=[LoadTypeCastInserter,
StreamTypeCastInserter], RULES_DISABLED=[AddForEach, ColumnMapKeyPrune,
ConstantCalculator, GroupByConstParallelSetter, LimitOptimizer, MergeFilter,
MergeForEach, PartitionFilterOptimizer, PredicatePushdownOptimizer,
PushDownForEachFlatten, PushUpFilter, SplitFilter]}
```

You will learn a few rules here.

Partition Filter Optimizer

Filter conditions immediately after the load can be pushed down to the `load` statements. HCatalog does this by loading the required partition data only if the partition filter is specified after the `load` statement.

The following code contains a filter that will be pushed to the loader statement:

```
Emp = load 'employee'  using HCatalogLoader as (eno,ename,salary,dno);
Dept100=filter emp by dno==100;
```

Merge foreach

If you have two `foreach` statements consecutively, this will merge two of them to have one `foreach` statement.

This will work if two are consecutive, the last one is not a nested `foreach`, and the first one does not contain `flatten`.

```
Pig latin code before rule
---------------------------
Emp = load '' using PigStorage() as (eno,ename,salary,dno);
Newsal=Foreach emp generate salary*1.1;
Nnewsal =foreach Newsal generate $0+10000;
```

```
Pig Latin code after Rule
-------------------------
Emp = load '' using PigStorage() as (eno,ename,salary,dno);
Newsal=Foreach emp generate salary*1.1+10000;
```

Constant Calculator

This will create a better expression for constants depending on user input.
Here is the Pig Latin code before the rule:

```
Sal= filter emp by salary>10000*10;
```

Here is the Pig Latin code after the rule is applied:

```
Sal= filter emp by salary>100000;
```

Pig provides many such rules. You can check the Pig documentation for the rest of the rules.

Cluster Optimization

After optimizing the Pig Latin script, you may still face performance issues if you do not optimize the underlying platform. The platform includes the best memory settings for Hadoop daemons, operating system settings, hard disk space, and Hadoop configuration.
You will learn a few things about the underlying platform here.

Disk Space

As MapReduce is disk-based, you will have a lot of disk operations. At least 20 percent of the disk space should be reserved for I/O operations. Otherwise, you will see a performance degradation on the cluster. It is a good idea to purge files periodically if they are not useful.
For example, log files can be removed periodically. You can write and schedule a shell script for file removal or you can configure log4j to do this.

Separate Setup for Zookeeper

The entire cluster will be dependent on Zookeeper (covered in Chapter 17). It is recommended that you have a dedicated disk for Zookeeper because it will have a lot of I/O operations, and it will also add value to the cluster if you can maintain a separate machine for Zookeeper.

Scheduler

There will be many cases where jobs fail or jobs do not get into running states because of a lack of resources on the cluster. You need to use a better scheduler configuration to avoid such cases.

The default scheduler is the capacity scheduler in most Hadoop distributions. Configure a queue for critical jobs and try to assign resources as much as possible.

The following properties set the minimum capacity for the analytics queue as 50 percent and the maximum capacity as 100 percent:

```
yarn.scheduler.capacity.analytics.maximum-capacity=100
yarn.scheduler.capacity.analytics.capacity =50
```

For more details about the capacity scheduler, you can check the Hadoop web site at http://hadoop.apache.org.

Name Node Heap Size

Many times the name node can go down depending on the number of files on the cluster.

One of reasons for the name node crash could be an insufficient heap size. You need to increase the heap size depending on the number of files on the cluster. The hdfs fsck command will display the number of files available on the cluster.

You need to change the HADOOP_NAMENODE_OPTS property value in hadoop-env.sh to increase the heap size of the name node. Table 16-1 lists the heap size for the name node against the millions of files in the cluster. For example, if you have 50 million files in the cluster, you need to have 24320 MB as the heap size.

Table 16-1. *How Heap Size Is Related to Number of Files*

number of files (millions)	Java heap size (MB)
<1	1024
01-05	3072
05-10	5376
10- 20	9984
20-30	14848
30-40	19456
40-50	24320
50-70	33536
70-100	47872
100-125	59648
125-150	71424
150-200	94976

Other Memory Settings

Many times you will see out-of-memory errors thrown by jobs and daemons. You can increase the memory if you see an out-of-memory error.

You need to understand how much memory you can increase, how much you should keep for the operating system, and, if you have HBase on the same cluster, how much memory you should keep for it. However, it is recommended that you have a separate cluster for HBase.

HBase and System Memory

Table 16-2 lists how much memory you should keep for HBase and the operating system.

Table 16-2. *Memory Constraints*

Memory per Node (GB)	Reserved memory for system (GB)	Reserved Memory for Hbase (GB)
4	1	1
8	2	1
16	2	2
24	4	4
48	6	8
64	8	8
72	8	8
96	12	16
128	24	24
256	32	32
512	64	64

For example, if you have a node memory of 64 GB, you can allocate 8 GB to the system and 8 GB for HBase.

The rest of the memory you can use for Hadoop daemons and jobs.

Container Memory

You should also fine-tune the container memory depending on the node memory as per Table 16-3.

Table 16-3. *Container Memory Related to RAM per Node*

RAM per node (GB)	Min. container size (MB)
<4	256
>4 and <8	512
>8 and <24	1024
>24	2048

For example, if the node memory is 24 GB, you can allocate a minimum of 2048 MB as the container size.

Increasing the container size will also improve the job performance but might decrease the total number of jobs the cluster can run.

Summary

In this chapter, you learned many optimization tips related to join strategies, best practices for writing Pig Latin code, implicit rules in Pig, and cluster optimization tips. The following are some important topics you learned:

- How to use replicated joins to avoid Reduce tasks in order to improve job performance

- How to achieve join functionality without the `join` operator by using the distributed cache

- How to avoid skewed data by implementing a skewed join

- How to implement a sort-merge join

- How to filter early, project early, and define the schema always

- How to store reusable data in a staging directory

- How to write the `Algebraic` function wherever applicable

- How to write the `Accumulator` function where `Algebraic` is not applicable

- How to compress intermediate data for better performance

- How to combine small inputs using `SequenceFileInputFormat`

- How to decide on the number of Reduce tasks at the script level and at the operator level

- How to enable performance-related counters for an UDF

- What optimization rules are enabled in Pig Latin

- How to decide on the memory size for the name node, HBase, and container

Hadoop Ecosystem Tools

In this chapter, you will learn the basics of some other Hadoop ecosystem tools such as Zookeeper, Cascading, Presto, Tez, and Spark.

Apache Zookeeper

Hadoop clusters can be scaled to thousands of nodes, and clusters can have multiple data-processing tools too. As most of these tools follow a master-slave approach, the slaves will be reporting to the master at regular intervals to signal they are alive. This communication needs to happen in real time.

Besides this simple reporting, there are several occasions where real-time communication needs to happen between nodes and services. Even a small delay in communication can cause failures in distributed applications.

Apache Zookeeper is a real-time coordination system for distributed applications. It is scalable, reliable, and tested in enterprise-level big data applications. It was originally developed by Yahoo, and now it is part of the Apache Software Foundation.

Zookeeper provides many critical services such as high availability, configuration management, and load sharing in big data clusters. In addition to these services, its general applications include group membership, leader election, and group messaging.

The Zookeeper service is the backbone to any big data cluster. Directly or indirectly, all technologies in big data clusters will depend on Zookeeper.

For example, Hadoop, Hive, Storm, Kafka, and HBase will directly depend on Zookeeper, and all Hadoop abstractions will indirectly depend on Zookeeper.

A production cluster needs to have separate machines, cooling, power cables, and switches for the Zookeeper service. It is also recommended that you have a dedicated disk for Zookeeper for better I/O performance.

Terminology

Zookeeper nodes are called *znodes*. One znode can have any number of children nodes, grandchildren nodes, great-grandchildren nodes, and so on.

Znodes are the same as directories in a file system. These nodes have data, versions, and some other statistics such as creation time, modification time, and so on.

© Balaswamy Vaddeman 2016
B. Vaddeman, *Beginning Apache Pig*, DOI 10.1007/978-1-4842-2337-6_17

Zookeeper has normal znodes, ephemeral znodes, and sequential znodes.

- Ephemeral nodes do not exist once a session is killed, and they do not have any children nodes.

- Sequential nodes have unique counters associated with their paths. Zookeeper adds a ten-digit number to a directory to uniquely identify it.

Zookeeper contains a group of servers called an *ensemble*. An ensemble needs to have $2*x+1$ servers, where x represents a number of failures that can survive. For example, if an ensemble has seven servers, it can survive three server failures. If more than three servers are down, the Zookeeper service will not be available.

Zookeeper follows a "leader-follower" philosophy that says one of the Zookeeper servers is the leader and the others are the followers. The leader will get first-hand information from the Zookeeper clients, and the followers will be updated by the leader. When you start the Zookeeper service, the first server will be elected as the leader, and the others will become the followers.

For any reason if the leader election has not happened, the Zookeeper service will not be available.

Zookeeper is a simple event-driven framework that can be used to develop powerful applications. It even has watcher functionality in which a client can be the watcher for a znode and Zookeeper will notify that client if any change is there in that znode.

Zookeeper uses this simple watcher functionality for most applications.

Applications

There are many Zookeeper applications. Some of them are group-related operations, configuration management, lock services, and leader election.

You will learn how some of the Zookeeper applications are implemented.

Group Membership

The children of a znode will be considered a group. The Zookeeper client can get children nodes of the Zookeeper node programmatically to perform group-related operations. When a znode gets deleted, it will become out of that group.

Configuration

Every znode can contain some data that can be used as a configuration. Multiple znodes can be created to store multiple configuration values. These values will become centralized, and clients will be notified when values are changed so that the client can perform the appropriate operations depending on the changes. Of course, clients need to be in the watchers list of a particular znode.

Lock Service

When multiple clients try to create a znode, the first client that succeeds in creating a znode will get a lock on that znode, and the other clients will become watchers of that znode.

When the lock holding the znode gets deleted, the znode watchers will be notified so that they can try creating the znode and get a lock on it.

Most of the Hadoop high availability features such as the name node and the resource manager are implemented using the Zookeeper lock service.

Open two terminals and connect to the Zookeeper CLI. In one terminal, create a znode called samplenode and make it the watcher using the get watch command.

In the other terminal, delete that znode, which will notify the client on the first terminal because it is the watcher of the znode sample.

Similarly, all watchers will be informed about the deletion of the znode so that they can try creating the same znode again to get the lock.

Command-Line Interface

You can start using the Zookeeper command-line interface using the zkCLi.sh shell script. It will be available under the bin folder of the Zookeeper home.

Figure 17-1 shows how to start the CLI of Zookeeper.

```
hdfs@cluster10:~> /usr/lib/zookeeper/bin/zkCli.sh
Connecting to localhost:2181
[zk: localhost:2181(CONNECTED) 0]
```

Figure 17-1. *Starting the CLI*

Once you have started the command-line interface, you can use the following commands to interact with Zookeeper. Use the help command to see a list of available commands.

stat

The stat command will display statistics such as creation time, modified time, and number of children of a znode. The command output will have statistics that are the same as the ls2 command.

ls

The ls command displays a list of available znodes on Zookeeper. The ls2 command also displays available znodes, including statistics creation time, modification time, number of children, and so on.

create

The create command creates a znode in Zookeeper. If data is specified, it also sets data. Specifying data is optional when creating a command. The -e option is used to create an ephemeral node, and the -s option is used to create a sequence node.

Figure 17-2 shows the create command usage.

```
[zk: localhost:2181(CONNECTED) 14] create /samplenode sampledata
Created /samplenode
[zk: localhost:2181(CONNECTED) 15] ▐
```

Figure 17-2. *The create command*

get

The get command retrieves the data of a znode and also displays statistics about the znode. You can even watch for data changes using the watch option.

The command in Figure 17-3 displays the data of samplenode and the client becomes the watcher for this znode.

```
[zk: localhost:2181(CONNECTED) 15] get /samplenode watch
sampledata
cZxid = 0x3b00010ee9
ctime = Sat Jul 23 10:14:40 EDT 2016
mZxid = 0x3b00010ee9
mtime = Sat Jul 23 10:14:40 EDT 2016
pZxid = 0x3b00010ee9
cversion = 0
dataversion = 0
aclversion = 0
ephemeralOwner = 0x0
dataLength = 10
numchildren = 0
```

Figure 17-3. *A sample get command*

set

The set command sets new data for a given znode, and the watchers will be notified if any are watching it.

delete

The delete command will delete the znode from Zookeeper, and also the watchers will be notified after the deletion.

Four-Letter Commands

Zookeeper provides four-letter commands that provide more information about the
Zookeeper service. All these commands will have four letters only. You will learn about
some four-letter commands here. All these commands need to be run using the Unix
telnet or nc program.

ruok

The ruok command is used to check whether a Zookeeper server is running. It returns
imok if it is running; otherwise, it returns nothing. It checks only the running state of
the server; if more information is required about the server, you need to use the srvr
command.

The following command uses the IP address and port number in the ruok command.
You can even use the host name instead of the IP address.

```
echo ruok | nc 127.0.0.1 2181
imok
```

srvr

The srvr command displays more details such as the number of connections, the node
count, and the mode of a server, as shown in Figure 17-4. The mode will inform you of
whether the server is a leader or a follower.

```
hdfs@cluster10-1:~> echo srvr|nc 127.0.0.1 2181
Zookeeper version: 3.4.6-3485--1, built on 12/16/2015 02:35 GMT
Latency min/avg/max: 0/0/61
Received: 65179
Sent: 65182
Connections: 11
Outstanding: 0
Zxid: 0x3b00017dbb
Mode: leader
Node count: 738
```

Figure 17-4. *Sample srvr command*

stat

The stat command displays statistics about a server. The command output will be the
same as the srvr command except the client details.

wchs

wchs displays a watchers list of a server. If you want a session-wise watchers list, you need to use whch, and if you want a path-wise watchers list, you need to use the wchp command.

Measuring Time

Time in Zookeeper is measured in *ticks*. You will define one tick value in milliseconds using the ticktime property. Based on the *tick time*, Zookeeper will have two more properties: the initial tick time and the sync tick time.

- The initial tick time is defined using the property init limit, which is the number of ticks required for a client to establish a connection with the Zookeeper server while it is initializing.

- The sync tick time is defined using the sync limit property, which is the number of ticks required for a client to establish a connection while the server is running.

Client connections will be refused if these times are exceeded.

With measuring time, you have completed the fundamentals of Apache Zookeeper. For more information about Apache Zookeeper, you can visit https://zookeeper. apache.org/.

Cascading

As discussed earlier in the book, Cascading is a Java-based MapReduce abstraction that is used to build data pipelines. It is an open source library written by Chris Wensel. Cascading uses plumbing terminology, so it is easy to correlate things. It can be used in production as it has already been adopted by companies like Twitter.

Here you will learn some basic features such as how to define the source and sink and how to use pipes and operations in Cascading. For more information, check its official web site at www.cascading.org/.

Defining a Source

You now will see how to define a source in Cascading.

Fields

The TextLine class is used to define fields for text data. It will read the offset and line of data the same as the output of the default Map task. Using TextLine, you can define both the source fields and the sink fields.

The following code defines the field `txtLine` for the source data:

```
TextLine txtLine = new TextLine( new Fields(new String[]{"txtline"})
```

The `TextDelimited` class is used to define the fields for delimited text files. You can specify a delimiter, and you can skip the header and the write header.

- When the skip header is specified, Cascading ignores the first line.

- When the write header is specified, it will write the header to the output files while generating output.

The following code creates the fields eno, ename, `salary`, and dno for the employee data set and asks Cascading to ignore headers and use a colon as a delimiter.

```
TextDelimited empfields=new TextDelimited(new Fields(new String[]{"eno",
"ename","salary","dno"}), true, ":");
```

Taps

Taps are responsible for both reading data from and writing data to the data location. Here you will create a source tap that is responsible for reading data and a sink tap that is responsible for writing data.

Some of the taps are discussed next.

HFS and LFS

The HFS tap is used to read data from and write data to the Hadoop distributed file system (HDFS), and the LFS tap is used to deal with the local file system data. These taps will take the schema and data set path.

The following code defines the source tap for the employee data set using its path:

```
TextDelimited empfields=new TextDelimited(new Fields(new String[]{"eno",
"ename","salary","dno"}), true, ":");
Tap sourceTap = new Hfs(empfields, "/data/emp" );
```

- Cascading supports JDBC; you can read data from RDBMS systems like Oracle, Derby, and MySQL. You can use the JDBC tap, and it can read data from a data warehouse system like Teradata.

- Cascading also provides taps for reading data from and writing data to NoSQL databases such as HBase, Accumulo, and Cassandra.

- Cascading can also access Hive table data.

You can get more information about other taps at `www.cascading.org/extensions/`.

Defining a Sink

Defining a sink is not very different from defining source. A Cascading sink will have sink mode information. A sink mode decides what to do with the existing output in the output path. Cascading provides three sink modes, as listed in Figure 17-5.

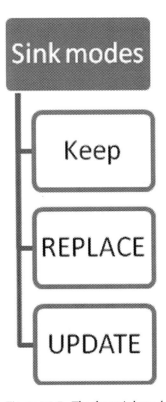

Figure 17-5. *The three sink modes*

- The default is the keep mode, which will fail application if the output directory exists. This is also the default behavior for any processing tools such as Pig and MapReduce.

- The replace mode will delete the output data and will write the newly generated output to the output path.

- The update mode will append data in the existing output directory.

The following code creates a sink tap that will replace the existing data in the output path with the newly generated output:

```
Fields empout= new Fields(new String[]{"dno" ,"count"});
Tap sinkTap = new Hfs(empout, "/data/emp",SINKMODE.REPLACE );
```

Pipes

In Cascading, the source tap and sink taps are connected by a collection of pipes. These pipes will allow you to perform the required operations on the data. Cascading provides pipes called each, every, group by, co-group, and subassembly.

- each will allow you to work on each row of data.

- every will allow you to work on every group key.

- group by is the same as a Pig Latin group that allows you to perform group operations.

- cogroup is the same as the Pig Latin cogroup operator that will allow you to perform joins.

- subassembly is the same as a Pig Latin macro that will allow you to create reusable pipes in Cascading.

The following code performs an inner join on two tables, emp and dept, using the common field dno:

```
CoGroup join=new CoGroup(emp, new Fields(new String[]{"dno"}), dept, new Fields(new String[]{"dno"}),new InnerJoin());
```

Types of Operations

Cascading allows you to perform four types of operations on its pipes, as listed in Figure 17-6.

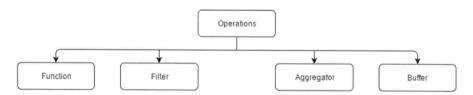

Figure 17-6. *The four operations*

Function

Function is a single-row function that performs the requested operation on a row and returns zero or more rows. You specify Function using each pipe. You can use built-in functions or you can use your own function.

You need to extend the BaseOperation class and implement the Function interface to write your own function. You need to call super within the function constructor. The super call specifies the number of incoming fields required and the output field name.

The following constructor says one incoming field is required, and the output field name is ename:

```
public class Upper extends BaseOperation implements Function {
public Upper(){
        super( 1, new Fields( new String[]{"ename"} ) );
}
```

You will implement the operate method to write the business logic. First you will take input fields using names or positions. You will write the necessary logic on those fields and will write the result into a tuple that will be returned to the user.

The following code takes one field and converts its value into uppercase:

```
public void operate(FlowProcess fproc, FunctionCall fcall) {
        TupleEntry args = fcall.getArguments();
        String ename =(String)args.get(0);
        String enameupper=ename.toUpperCase();
        Tuple output = new Tuple();
        output.addString(enameupper);
        fcall.getOutputCollector().add(output);

}
```

The following program uses a custom function to convert employee names into uppercase. This program can also be run using Eclipse as it is using Filetap.

```
public class UpperCase {


    public static void main(String[] args) {
        Tap srcTap = new FileTap(new TextDelimited(new Fields(new String[] {
                "eno", "ename", "salary", "dno" }), true, ","), "employee.csv");
        Tap sinkTap = new FileTap(new TextDelimited(new Fields(
                new String[] { "ename" })), "uppercase");
        Pipe emp = new Pipe("emp");
//will take only one field ename and perfrom upper case operation
        Pipe ename = new Each(emp, new Fields(new String[] { "ename" }),
                new Upper());
        Properties properties = new Properties();
        AppProps.setApplicationJarClass(properties, WordCount.class);
        LocalFlowConnector flowConnector = new LocalFlowConnector();
        Flow flow = flowConnector
                .connect("convertcase", srcTap, sinkTap, ename);
        flow.complete();

    }

}
```

Filter

Filter is same as the Pig Latin filter operator that performs given conditions on each row.

The following code performs a limit filter and returns two rows:

```
public class LimitN {

    public static void main(String[] args) {
        Tap srcTap = new FileTap( new TextDelimited( new Fields(new String[]
                    {"eno","ename","salary","dno"}),true,",") , "employee.csv");
        Tap sinkTap = new FileTap( new TextDelimited( new Fields(new String[]
                    {"eno","ename","salary","dno"})) , "employeelimit");
        Pipe emp=new Pipe("emp");
        Pipe ename=new Each(emp, new Limit(2));
        Properties properties = new Properties();
          AppProps.setApplicationJarClass( properties, LimitN.class );
          LocalFlowConnector flowConnector = new LocalFlowConnector();
          Flow flow = flowConnector.connect( "limit", srcTap, sinkTap,ename );
          flow.complete();

    }

}
```

You can write your own filter by extending BaseOperation and implementing the filter interface. You will implement the isRemove method, which will return a Boolean value.

The following code removes the salary field with a null value:

```
public class SalryFilter extends BaseOperation implements Filter {

    public boolean isRemove(FlowProcess fproc, FilterCall fcall) {
        TupleEntry emptuple=fcall.getArguments();
        return null==emptuple.get(0);
    }

}
```

Aggregator

Aggregator is used to write multirow functions that perform group-level operations. Aggregators can be used only with every pipe and cannot be used with each pipe.

You can write your own aggregator extending BaseOperation and implementing the Aggregator interface. You need to implement three methods: start, aggregate, and complete.

- The start method will have the initialization logic.

- The aggregate method will have the logic for the group operations.

- The complete method will be called in the end to perform cleanup operations.

The following code contains the structure for the custom aggregator class:

```
public class SampleAggregator extends BaseOperation implements Aggregator {

public void start(FlowProcess fproc, AggregatorCall aCall) {
    }

    public void aggregate(FlowProcess fproc, AggregatorCall aCall) {
    }

    public void complete(FlowProcess fproc, AggregatorCall aCall) {
    }

}
```

Buffer

A buffer is the same as an aggregator that can be used to perform group-related operations. Buffer provides an extra facility to iterate input tuples. This is useful for performing operations such as ranking.

You can write your own buffer by extending BaseOperation and implementing the Buffer interface. You need to implement the operate method from the Buffer interface.

The following code contains the structure for a custom buffer class:

```
public class SampleBuffer extends BaseOperation implements Buffer {

    public void operate(FlowProcess fproc, BufferCall bcall) {
    }

}
```

With this type of operation, you have completed the fundamentals of Cascading. For more information about Cascading, visit www.cascading.org/.

Apache Spark

Apache Spark is in-memory parallel computing framework used for Internet-scale data processing. It was first developed at the University of California – Berkeley, and now it is part of the Apache Software Foundation.

As MapReduce is a disk-based data processing framework, it does not provide good I/O performance, and it is suitable only for batch-processing jobs. Complementing MapReduce, Spark provides in-memory computing and makes it suitable for interactive data analysis.

Spark is mostly used in these two use cases:

- Iterative data analysis requires performing multiple operations on the same data set until the required output is produced. Many data science use cases are in this category.

- Interactive data analysis allows you to perform data analysis on demand like you do using RDBMSs.

Apache Spark introduces a parallel computing paradigm called *Resilient Distributed Datasets* (RDD). RDD allows you to perform in-memory computing in parallel and in a fault-tolerant manner.

Spark provides five modules called Core, Spark SQL, Streaming, MLlib, and GraphX, as listed in Figure 17-7.

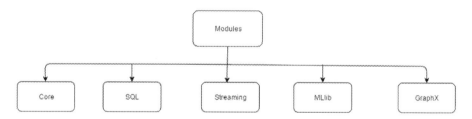

Figure 17-7. *The five modules of Spark*

- Core is the base module that is required by all the other modules.

- Spark SQL provides the SQL interface to process structured data.

- Streaming is used to process streaming data.

- MLlib is a machine learning library from Spark. Spark MLlib provides high-performant and scalable machine learning algorithms and supports R and Python. It can also integrate with NumPy. As a machine learning algorithm requires iterative computation, MLlib is most suitable for it.

- GraphX is a graph processing framework developed in Spark. A use case of GraphX is the same as Apache Giraph that does graph processing. The only difference is that Giraph is based on MapReduce and involves disk operations, and GraphX does in-memory graph processing and will be faster.

Apache Spark supports NoSQL databases, warehouse systems such as Hive, RDBMSs such as Oracle, and flat files.

You can write Spark applications in programming languages such as Java, Spark, and Python. Here you will learn how to write Java applications.

Core

Core is a base module that provides a platform for all other modules in Spark. It provides an API that includes RDD operations, scheduling, and task completion.

Configuration

You can create a configuration object for Spark using the SparkConf class. The SparkConf class provides setter methods for setting the property value and getter methods for retrieving property values within the Spark program.

You can set a master using the setMaster method. You can use local to run the program in the local file system and local[number of core] to run the program in local mode using the given number of cores.

If you want to run the program in cluster mode, you need to set the master to spark://masternode:portnumber.

You can retrieve all the property values using the getAll method or you can retrieve the value of a particular property using the get method.

The following code sets the master as local to run this program in local mode, sets the application name as WordCount, and sets the executor memory as 8 GB.

```
SparkConf sparkConf = new SparkConf().setAppName("WordCount")
      .setMaster("local")
      .set("spark.executor.memory", "8g");
```

JavaSparkContext enables you to perform all the computing operations. You need to create an object of it before calling RDD operations. You require the sparkconf object to create JavaSparkContext. There will be only one Spark context per JVM.

The following code instantiates JavaSparkContext:

```
JavaSparkContext ctx = new JavaSparkContext(sparkConf);
```

Sourcing

You can specify the input path using the JavaSparkContext object. The text method returns the RDD of each record for the input text files. The wholeTextFiles method returns the RDD of the key-value pair where the key is the file path and the value is the data of the file. This is used for small files.

You can use the binaryFiles method to read binary files and the sequenceFiles method to read Hadoop sequence files. All these methods take the file path URI, which can refer both to the local file system and to the Hadoop file system.

The following code looks for the input path in the local file system if the set master is local in the Spark configuration; otherwise, it looks for it in the Hadoop distributed file system:

```
JavaRDD<String> lines = javaSparkContext.textFile("data\\in");
```

You can also read a file from Hadoop using a custom input file format. A specified input format class needs to be available in the class path.

The following code specifies the HDFS file path:

```
javaSparkContext.hadoopFile("data\\in",FileInputFormat.class, LongWritable.
class, Text.class);
```

Specifying a Sink

Once you have required the output in RDD, you will write output to persistence storage. You can write the output file as the text file, object file, and Hadoop file with a custom format using methods in the JavaRDD class. You can also specify a codec while writing data to the output directory.

The following code writes the output as the text file using the Gzip codec:

```
counts.saveAsTextFile("data\\wordcountout",GzipCodec.class);
```

Operations

All RDD operations are of two types: transformations and actions. Transformations will transform data, and actions are metrics to be computed such as count, sum, and average.

For example, all map functions are transformations, and all reduce functions are actions. Transformations are lazy, and they are not run until actions are called.

You can write all map functions and reduce functions as inline functions or you can use lambda expressions.

The following code reads the employee name from the employee file containing the employee number, employee name, salary, and department number and converts it into uppercase:

```
JavaRDD<String> lineText = javaSparkContext.textFile("data\\employee");
    JavaRDD<String> uppername=lineText.map(new Function<String, String>() {

            @Override
            public String call(String str) throws Exception {

                String[] words=str.split(",");
                String uppername=words[1].toUpperCase();
                return uppername;
        }
    });
```

You can also write the function in a separate class file and refer to that function within the map function. You need to implement the call method of the Function interface, as shown here:

```
public class FunUpperCase implements Function<String, String> {

    @Override
    public String call(String str) throws Exception {
        String[] empdetails = str.split(",");
        return empdetails[1].toUpperCase();
    }

}
```

This function will be called once per every line. You need to instantiate it within a map function to invoke it, as shown here:

```
JavaRDD<String> uppername = lineText.map(new FunUpperCase());
```

Similarly, you can use existing reduce functions, and you can also write your own reduce functions in Spark.

SQL

As discussed earlier, Spark provides a module called Spark SQL that provides the SQL interface for processing structured data.

Spark SQL provides SQL support for the Hadoop-based warehouse Apache Hive, RDBMSs like Oracle and Postgres using JDBC, and also file system data.

You can use programmatic SQL and also traditional SQL. You need to create an object for SQLConntext for SQL support.

```
JavaSparkContext javaSparkContext = new JavaSparkContext(sparkConf);
SQLContext sqlContext = new SQLContext(javaSparkContext);
```

Spark SQL contains two concepts called data sets and data frames.

- *Data sets* are the same as in RDD that represent data sets with better serialization.

- *Data frames* are similar to RDBMS tables that have column names.

Source

You can provide the source using the DataFrameReader object. You can use the table method to specify the table as a source, the jdbc method to specify the RDBMS table, and the text method to specify the text data set as a source.

You can specify ORC, Parquet, and JSON format files as a source using the orc, parquet, and json methods.

The following are some examples for specifying the source. The read method on sqlContext returns DataFrameReader.

The following code specifies the text data set as the source:

```
DataFrame df = sqlContext.read().text("data\\employee*");
```

The following code specifies the orc data format as the source:

```
DataFrame df = sqlContext.read().orc("data\\employee*");
```

The following code specifies the table from a Postgres database as the source:

```
DataFrame df = sqlContext.read().jdbc("jdbc:postgresql://localhost:5432/
empdb","employee");
```

Data-Processing Methods of Data Frames

Data frames provide many useful methods to perform data analysis operations. You will learn about some of them here.

The show method displays records to the console, and you can also specify the number of records to display.

The following code displays ten records from the source:

```
JavaSparkContext javaSparkContext = new JavaSparkContext(sparkConf);
SQLContext sqlContext = new SQLContext(javaSparkContext);
DataFrame dFrames = sqlContext.read().text("data\\in\\p*");
dFrames.show(10);
```

The printSchema() method works the same as the describe operator of Pig Latin. The following code displays the default schema that is a single-column table:

```
dFrames.printSchema();
```

```
root
 |-- value: string (nullable = true)
```

You can specify conditions on a column like with the filter operator of Pig Latin using the filter method.

The following code looks for the word *bala* in a column named value and displays some 20 records. By default, the show method displays 20 records.

```
df.filter("value like '%bala%' ").show();
```

Spark SQL is intuitive, and it does not take much time to write applications.

The following code prints the number of unique lines in a data set. This program sets a directory with the bin\winutils.exe file as the Hadoop home directory so that it can run within the Windows OS also.

```java
public class FramesDemo {
    public static void main(String[] args) {
        System.setProperty("hadoop.home.dir", "hadoop-winutils");
            SparkConf sparkConf = new SparkConf().setAppName("FramesDemo")
                .setMaster("local").set("spark.executor.memory", "2g");
        sparkConf.set("spark.io.compression.codec",
                "org.apache.spark.io.LZ4CompressionCodec");
        JavaSparkContext javaSparkContext = new JavaSparkContext(sparkConf);
        SQLContext sqlContext = new SQLContext(javaSparkContext);
        DataFrame dFrames = sqlContext.read().text("data\\emp\\*");
    long rows = dFrames.orderBy("value").distinct().count();
        System.out.println(rows);
        javaSparkContext.stop();
    }

}
```

Running SQL Queries

You can also run SQL queries using Spark SQL. You need to register the input data as a temporary table before running the SQL queries.

The following code displays the schema of the just-registered table named temp:

```java
SQLContext sqlContext = new SQLContext(javaSparkContext);
DataFrame dFrames = sqlContext.read().text("data\\emp\\*");
dFrames.registerTempTable("temp");
DataFrame schema=sqlContext.sql("describe temp");
schema.show();
```

```
+--------+---------+-------+
|col_name|data_type|comment|
+--------+---------+-------+
|   value|   string|       |
+--------+---------+-------+
```

You can even specify the schema whatever you want with respect to the data format at run time.

1. You need to write a Plain Old Java Object (POJO) class that contains setter and getter methods for all the columns. This needs to implement the serializable interface so that data can be transferred over the network.

The following code contains the Employee data set schema:

```
public class Employee implements Serializable {
      private int eno;
    private String ename;
    private int salary;
    private int dno;
      public int getEno() {
        return eno;
    }
      public void setEno(int eno) {
        this.eno = eno;
    }
      public String getEname() {
        return ename;
    }
      public void setEname(String ename) {
        this.ename = ename;
    }
      public int getSalary() {
        return salary;
    }
    public void setSalary(int salary) {
        this.salary = salary;
    }
      public int getDno() {
        return dno;
    }
      public void setDno(int dno) {
        this.dno = dno;
    }

}
```

2. In the next step, you need to call all the setter methods inside a map function, and it should return the JavaRDD of the Employee record.

    The following code applies the schema using the Employee POJO class methods:

```
JavaRDD<Employee> emp=javaSparkContext.textFile("data\\in\\
emp*").map(new Function<String, Employee>() {

        @Override
        public Employee call(String str) throws Exception {
            String[] erecord=str.split(",");
            Employee emp=new Employee();
```

```
                        emp.setEno(Integer.parseInt(erecord[0]));
                        emp.setEname(erecord[1]);
                        emp.setSalary(Integer.parseInt(erecord[2]));
                        emp.setSalary(Integer.parseInt(erecord[3]));

                        return emp;
                }
        });
```

3. Now you need to convert JavaRDD into DataFrame, applying the schema of the employee POJO class as shown here:

```
DataFrame empSchema = sqlContext.createDataFrame(emp, Employee.class);
```

4. As a final step, you need to register a temp table before applying the SQL queries, as follows:

```
empSchema.registerTempTable("employee");
```

Now, you can run the SQL queries on the newly registered table employee. The following code returns the schema of the new employee table.

```
sqlContext.sql("describe employee").show();
```

```
+--------+---------+-------+
|col_name|data_type|comment|
+--------+---------+-------+
|     dno|      int|       |
|   ename|   string|       |
|     eno|      int|       |
|  salary|      int|       |
+--------+---------+-------+
```

The following code returns the employee number and employee name of the employee table whose name is bala.

```
sqlContext.sql("select eno,ename from employee where ename like '%bala%'").
show();
+---+-----+
|eno|ename|
+---+-----+
|  1| bala|
+---+-----+
```

You have completed the learning fundamentals about Spark Core and SQL. For more information about Apache Spark, visit http://spark.apache.org/.

Apache Tez

MapReduce is a disk-based framework that involves a lot of I/O operations, so it can support only batch-processing jobs. MapReduce performance bottlenecks are addressed by Apache Tez.

Apache Tez is an execution engine that enables interactive data processing in the Hadoop ecosystem.

In MapReduce, the Map task is mandatory before the Reduce task. Particularly when you have multiple MapReduce jobs to be performed in a row, there will be many unnecessary Map tasks involved, and there will be a lot of I/O operations involved as the map output data is written to disk.

Figure 17-8 starts Map task map2 after completing Reduce task reduce1.

Figure 17-8. *The processing pipeline*

Tez will reduce the number of Map tasks wherever necessary to improve job performance. Tez also provides better resource management by reusing Hadoop containers.

Tez is built on top of YARN. It does not modify any data-processing logic. It only optimizes data processing.

Client applications such as Hive, Pig, and Cascading can use Tez to improve application performance.

As discussed earlier, you can choose Tez as the execution engine in Pig Latin using the -x option, as shown here:

```
pig -x tez pigscript.pig
```

In Apache Hive, you can choose Tez using the hive.execution.engine property, as shown here:

```
set hive.execution.engine=hive;
```

Tez supports both cluster mode and local mode. In cluster mode, it can access and process HDFS data, and in local mode it can process local file system data. However, only the client application will choose between the local file system and the Hadoop file system.

For more information about Apache Tez, visit http://tez.apache.org.

Presto

Enterprises will have their data in multiple technologies. For example, some data may be in RDBMSs, some might be in NoSQL databases, and some might be in file systems such as HDFS. If you want to get a report from more than one source, you need to move all the data to one platform so that you can generate the report.

Moving data to a different technology is a costly operation because you need to maintain separate storage for it. Also, moving data takes time and cannot provide ad hoc reporting. This problem is addressed by a product called Presto.

Presto is a parallel computing technology used as a SQL-based query engine that allows you to perform interactive data analysis in a multisource environment.

Presto primarily provides two benefits.

- You need not move data to a common platform for data analysis.

- Presto launches low-latency jobs to provide faster results.

As it provides support for standard ANSI SQL, it is easy to adopt. It is an open source product initially developed by Facebook; it has been tested on petabytes of data. Several other companies such as Twitter, LinkedIn, Netflix, and Uber have also tested it in production. Teradata is actively contributing to Presto and also provides commercial support.

- Presto requires Linux or Mac OS, 64-bit Java 8, and Python 2.4 or above.

- Presto is written in Java and can be installed on any Hadoop cluster.

- Presto is not a MapReduce abstraction.

Architecture

Presto follows a master-slave architecture that contains two types of nodes: coordinator and worker.

Figure 17-9 shows the Presto architecture.

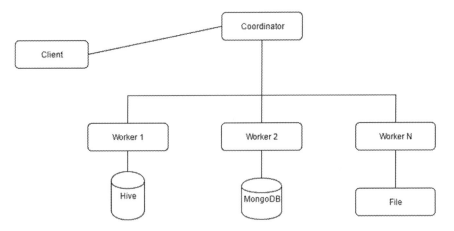

Figure 17-9. The Presto architecture

- A *coordinator* node is responsible for query parsing, analyzing, and planning. Once planning is done, it schedules the query on one of the closest worker nodes and monitors the progress.

- A *worker* node is responsible for connecting to the source system and retrieving the results from it and also for data processing.

Connectors

Presto provides connectors for RDBMSs such as MySQL and Postgres; for NoSQL databases such as Redis, MongoDB, and Cassandra; and for Apache Hive.

You need to define their connection properties in the `.properties` file under the `etc/catalog` folder.

Presto also supports business intelligence tools like Tableau and QlikView.

Pushdown Operations

Presto can push down predicates to the source system so that it has to fetch less data to the worker node. For example, say you have a MySQL query to count the number of employees in a department named IT, as shown here:

```
Select  count(eno) from employee where dname='IT';
```

Presto retrieves only the IT department records to the worker node by pushing the condition to the source system MySQL. The worker node performs the count operation to provide the final result.

For more information, please visit the Presto official web site at `https://prestodb.io/`.

Summary

In this chapter, you learned the fundamentals of the Hadoop ecosystem tools Zookeeper, Cascading, Apache Spark, Apache Tez, and Tez.

- You learned to use Zookeeper CLI commands such as `get`, `create`, `ls`, and so on.

- You learned to use Zookeeper services such as group membership, configuration management, and lock services.

- You learned to use the four-letter commands of Zookeeper such as `srvr`, `stat`, `ruok`, and so on.

- You learned to define the Cascading source and sink using HFS and LFS.

- You learned to use Cascading pipes such as `each`, `every`, `group`, `cogroup`, and `subassembly`.

- You learned to use Cascading operations such as `Function`, `Filter`, `Aggregator`, and `Buffer`.

- You learned that Apache Spark contains five modules. They are Core, SQL, Streaming, MLlib, and GraphX.

- You learned about configuration management in Apache Spark.

- You learned that Spark's RDD operations are of two types: transformations and actions.

- You learned how to run SQL queries on a file data source using data frames.

- You learned that Apache Tez is an execution engine built on top of YARN to address the performance issues of MapReduce.

- You learned that Presto is a parallel computing technology used as a SQL-based query engine that allows you to perform interactive data analysis in a multisource environment.

- You learned that Presto contains two types of nodes: coordinator and worker.

- You learned that Presto provides connectors for MySQL, Postgres, Redis, MongoDB, Cassandra, Apache Hive, and so on.

- You learned that Presto can push down predicates to source systems.

APPENDIX A

Built-in Functions

In this appendix, you will learn short definitions of built-in functions in Pig Latin. Examples are included wherever necessary. The functions are sorted alphabetically.

ABS
Returns the absolute value of a given input value

AccumuloStorage
Function for both loading and storing data from/to Accumulo

ACOS
Returns the arc cosine value of the given input

AddDuration
Returns the date after adding a specified duration to the input date

ARITY
Returns the number of fields in the input tuple

Syntax:
ARITY(exp)

Eaxmple:

```
foreach emp generate ARITY(emptuple);
```

ASIN
Returns the arc sin value of a given input

ATAN
Returns the arc tan value of a given input

AVG
Computes the average value of a given input field for an available group in script

Syntax:
AVG(exp)

Example:

```
grp = group emp all;
Avgsalary = foreach grp generate AVG(emp.salary);
```

AvroStorage
Function for both loading and storing Avro data files

BagToString
Converts bag data to chararray data using the delimiter specified

BagToTuple
Converts a bag to a tuple with respect to its structure

BinStorage
Function for both loading and storing binary data

Bloom
Builds a small size of data for resource-intensive operations like joins

CBRT
Returns the cube root of the given input value

CEIL
Returns the next greater integer

CONCAT
Performs the concatenation operation on input fields

COR
Returns the correlation for input data sets

COS
Returns the cosing value of the given input

COSH
Returns the hyperbolic cosine value of the given input

COUNT
Calculates the count of rows in a group

COUNT_STAR
Calculates the count of all rows including null in a group

COV
Returns the covariance for the input data sets

CurrentTime
Returns the current time

DaysBetween
Returns the difference of two dates in days

DIFF
Computes the difference of two bags and returns tuples that are present in only one bag

ENDSWITH
Checks whether one string ends with another string and returns a Boolean value

EqualsIgnoreCase
Checks whether two strings are equal irrespective of case

EXP
Computes Euler's number *e* raised to the power of the given input number

FLOOR
Returns the output of the mathematical operation FLOOR

GetDay
Returns the day from the input date

GetHour
Returns the hour from the input date

GetMilliSecond
Returns only milliseconds

GetMinute
Returns minutes from the input date

GetMonth
Returns the month from the input date
Example:

```
mon = foreach dummy generate GetMonth(ToDate('12-04-2016','dd-MM-yyyy'));
```

Output:

(4)

GetSecond
Returns the second from the input date

GetWeek
Returns the week from the input date

GetWeekYear
Returns the week number of year from the input date

GetYear
Returns the year from the input date

HBaseStorage
Function for both loading and storing HBase tables

HiveUDAF
Allows you to use Hive UDAFs in Pig Latin scripts

Example:

```
define hivesum HiveUDAF('sum');
 emp = load  'employee' using PigStorage(',')  as  (eno:int,ename:chararray,
salary:double,dno:int);
dnogrp = group emp by dno;
dnosum = foreach dnogrp generate emp.dno,hivesum(emp.salary);
```

HiveUDF
Allows you to use Hive UDFs in Pig Latin scripts

HiveUDTF
Allows you to use Hive UDTFs in Pig Latin scripts

HoursBetween
Returns the difference of two dates in hours

INDEXOF
Returns the position of a string within another string

IsEmpty
Checks whether the input bag or tuple is empty

JsonLoader
Function for loading JSON files

JsonStorage
Function for storing JSON files

KEYSET
Returns the key set from the input map data

Example:
Input file:

```
hdfs@cluster10-11:~> cat empmap.csv
[100#bala],100000,121
[101#Nirupam],120000,122
[102#radha],200000,123
[103#Nitya],1600000,124
```

Pig Latin code:

```
emp = load 'empmap.csv' using PigStorage(',') as (empmap:MAP[],sal,dno);
emp = foreach emp generate KEYSET(empmap);
dump emp;
```

Output:

```
({(100)})
({(101)})
({(102)})
({(103)})
```

LAST_INDEX_OF
Returns the last position of a string within another string

LCFIRST
Returns the input string after converting the first character into lowercase

LOG
Returns the logarithm value of the given input

LOG10
Returns the logarithm value of base 10 for the given input

LOWER
Returns lowercase for the input string

LTRIM
Removes the left-side spaces of a string

MAX
Computes the maximum value from a group

MilliSecondsBetween
Returns the difference of two dates in milliseconds

MIN
Computes the minimum value from a group

MinutesBetween
Returns the difference of two dates in hours

MonthsBetween
Returns the difference of two dates in months

Example:

```
mondiff = foreach dummy generate MonthsBetween(ToDate('12-08-2016',
'dd-MM-yyyy'),ToDate('12-07-2016','dd-MM-yyyy'));
```

Output:

(1)

OrcStorage
Function for both loading and storing ORC format files

ParquetLoader
Load function used for loading Parquet data

ParquetStorer
Store function used for storing Parquet data

PigDump
Stores data in UTF-8 format

PigStorage
Default function for both loading and storing data sets

PluckTuple
Allows you to select required fields with RegEx

RANDOM
Returns random number between 0.0 <= RANDOM < 1.0

REGEX_EXTRACT
Returns string matched with the RegEx and position

REGEX_EXTRACT_ALL
Returns the strings matched with the RegEx

REPLACE
Replaces one string with another

ROUND
Returns the result of the mathematical round function

ROUND_TO
Returns the result of the mathematical round function after rounding to a specified number of digits

RTRIM
Removes the right-side spaces

SecondsBetween
Returns the difference of two dates in seconds

SIN
Returns the sine value of the given input

SINH
Returns the hyperbolic sine value of the given input

SIZE
Returns the size of the argument depending on its data type (for example, bag returns the number of tuples and bytearray returns the number of bytes)

SPRINTF
Converts the given string into the required format

SQRT
Returns the square root of the given number

STARTSWITH
Checks whether the first string starts with the second string

STRSPLIT
Splits the string using the RegEx

STRSPLITTOBAG
Returns a bag of strings after splitting the input string

SUBSTRING
Returns a substring from the input string

SUBTRACT
Returns the output of Bag1-Bag2; minus is a set operation

SubtractDuration
Returns the date after substracting the specified duration from the input date

SUM
Computes the sum of a field for a group

TAN
Returns the tan value of the given input

Example:

```
tan90 = foreach dummy generate TAN(90);
```

Output:

```
(-1.995200412208242)
```

TANH
Returns the hyperbolic tan value of input

TextLoader
Load function for unstructured data

TOBAG
Converts the given input to a bag

ToDate
Converts the given input to a date

TOKENIZE
Breaks the given input into words or characters called *tokens* using the specified delimiter word or character

TOMAP
Converts the given input to a map

ToMilliSeconds
Converts the given input to milliseconds

TOP
Returns the top *n* rows

ToString
Converts the given date input to a string

TOTUPLE
Converts the given inputs to a tuple

ToUnixTime
Returns the Unix epoch time

TrevniStorage
Function for both loading and storing Trevni files

TRIM
Removes the spaces on both sides

UCFIRST
Returns the string in uppercase

UniqueID
Returns the unique ID for every record based on the task ID

UPPER
Returns an uppercase string

VALUELIST
Returns a list of values from the input map data that does not contain any duplicate values

VALUESET
Returns a set of values from the input map data that does not contain any duplicate values

WeeksBetween
Returns the difference of two dates in weeks

YearsBetween
Returns the difference of two dates in years

APPENDIX B

Apache Pig in Apache Ambari

Manually installing a Hadoop cluster and maintaining it is a tedious task. This problem is addressed by products such as Cloudera Manager and Apache Ambari.

- Cloudera Manager is proprietary software from the Cloudera company.

- Apache Ambari was developed by Hortonworks. Now it is an open source product that belongs to the Apache Software Foundation.

Apache Ambari is a web application used for cluster provisioning, configuration management, and monitoring the Hadoop cluster.

In this chapter, you will learn how to use Pig in Apache Ambari.

A Pig page in Ambari displays Summary, Configs, and Service Action options in the Ambari UI, as shown in Figure B-1. The Summary tab displays the number of Pig clients installed on the cluster. If you click Summary, it will display all the nodes where the Pig clients are installed.

Figure B-1. *The Pig page*

The Configs tab displays all the properties defined in the `pig.properties`, `pig-env.sh`, and `log4j.properties` files.

You can modify Pig properties through this page in Ambari.

Modifying Properties

If you modify a property in a required section and hit Save, Ambari 2.3 will ask you to name the configuration changes. Once they are saved, Ambari will ask you to restart the affected components so that the changes get applied.

Figure B-2 asks the user to restart seven components.

Figure B-2. *Restart prompt*

For example, the exectype value is set to mapreduce by default. You can change it to tez in the Advanced Pig Properties section, as shown here:

exectype=tez

Once you make any changes, Ambari will save its version; later you can just click it to revert the changes.

Service Check

You can check the working status of Pig using the service check feature in the Ambari user interface. If you click Run Service Check under Service Actions, Ambari will launch a new Pig job and will check the status of that job.

If a job has failed, Ambari will fail the service check for Pig. It will also display an error; you can check it from the OPS area in Ambari.

Most of the time a service check will fail if there is an issue with Apache Hadoop.

If the job succeeds, Ambari displays that the service check is successful, which means Pig is in working status.

Figure B-3 shows that the service check for Pig has failed.

0 Background Operations Running   X

Operations	Start Time	Duration	Show:	All (10)	▾
🔴 Pig Service Check	Today 19:01	120.42 secs	▓▓▓▓▓▓▓▓	100% ▶	

Figure B-3. *Failure of the Pig service*

Installing Pig

Manually installing Pig is a tedious task because you need to do it once per node. That is, if you have a 100-node cluster, you need to do it 100 times.

Apache Ambari simplifies this by automating this setup process with a user-friendly web interface.

You can install Pig using Ambari if it is not installed yet. You can click the Add Service option under the Actions drop-down, which will display the technologies available for installation. You can choose Pig and click Next. The rest of the steps are intuitive and can easily be performed.

Pig Status

Ambari provides the REST API. You can check the status of Pig by using it. You need the IP address where Ambari is running, the port number of Ambari, and the Hadoop cluster name.

Here's the syntax:

```
http://<IPaddress>:<portnumber>/api/v1/clusters/<clustername>/services/
PIG?fields=ServiceInfo/state
```

You can use this URL in a web browser, or you can use the curl command in a Unix terminal. This will return output in JSON format, as shown here:

```
http://112.11.11.100:8080/api/v1/clusters/Cluster10/services/
PIG?fields=ServiceInfo/state

{
  "href" : "http://112.11.11.100:8080/api/v1/Cluster10/Cluster10/services/
PIG?fields=ServiceInfo/state",
  "ServiceInfo" : {
    "cluster_name" : "Cluster10",
    "service_name" : "PIG",
    "state" : "INSTALLED"
  }
}
```

This output says Pig is installed.

If you do not know the available Hadoop clusters, you can check using the clusters option, as shown here, which will display the cluster names:

```
http://112.11.11.100:8080/api/v1/clusters/
```

Check All Available Services

You can even get all the available services from a cluster using the services option, which will also include Pig if installed.

The following code displays the available services in a cluster named cluster10:

`http://112.11.11.100:8080/api/v1/clusters/Cluster10/services`

The following is the partial output:

```
{
    {
      "href" : "http://112.11.11.100:8080/api/v1/clusters/Cluster10/
                services/HDFS",
      "ServiceInfo" : {
        "cluster_name" : "Cluster10",
        "service_name" : "HDFS"
      }
    },
    {
      "href" : "http://112.11.11.100:8080/api/v1/clusters/Cluster10/
                services/PIG",
      "ServiceInfo" : {
        "cluster_name" : "Cluster10",
        "service_name" : "PIG"
      }
    },
    .
    .
    .
    {
      "href" : "http://112.11.11.100:8080/api/v1/clusters/Cluster10/
                services/SPARK",
      "ServiceInfo" : {
        "cluster_name" : "Cluster10",
        "service_name" : "SPARK"
      }
    }
  ]
}
```

Summary

In this appendix, you learned about the Pig features in Apache Ambari, including the following:

- How to change the Pig configuration using Ambari

- How to check the working status of Pig using a service check

- How to use the Ambari REST API to see all the installed services in a cluster

APPENDIX C

HBaseStorage and ORCStorage Options

In this appendix, you will learn about the HBaseStorage and ORCStorage options.

HBaseStorage

The HBaseStorage function allows you to specify three types of conditions. They are row-based conditions, timestamp-based conditions, and other types of conditions.

Row-Based Conditions

The following loads the row key as the first value in every tuple returned from HBase. Its default value is false.

```
-loadKey=(true|false)
```

The following allows you to specify a greater-than condition on the row key. It retrieves a tuple whose row is greater than the specified key value.

```
emp = load 'hbase://employee' using org.apache.pig.backend.hadoop.hbase.
    HBaseStorage('empdetails:*','-loadKey=true');
dump emp;
-gt=KeyValue
```

The following allows you to specify a less-than condition on the row key. It retrieves a tuple whose row is less than the specified key value.

```
-lt=KeyValue
```

© Balaswamy Vaddeman 2016
B. Vaddeman, *Beginning Apache Pig*, DOI 10.1007/978-1-4842-2337-6

The following allows you to specify regular expressions on the row key. It retrieves a tuple whose row matches with the specified RegEx.

```
-regex=regexvalue
```

The following allows you to specify a greater-than or equals condition on the row key. It retrieves a tuple whose row is greater than or equal to the specified key value.

```
-gte=KeyValue
```

The following allows you to specify a less-than or equals condition on the row key. It retrieves a tuple whose row is less than or equal to the specified key value.

```
-lte=KeyValue
```

Timestamp-Based Conditions

The following will allow you to specify a greater-than or equal condition on the timestamp. It will return values whose timestamps are greater than or equal to the specified timestamp.

```
-minTimestamp=timestamp
```

The following example will allow you to specify a less-than or equal condition on the timestamp. It will return values whose timestamps are less than or equal to the specified timestamp.

```
emp = load 'hbase://emp' using org.apache.pig.backend.hadoop.hbase.
HBaseStorage('empdetails:*','-loadKey=true -minTimestamp=1466922018861') ;
dump emp;
-maxTimestamp=timestamp
```

This will allow you to specify the equal condition on the timestamp. It will return values whose timestamps equal the specified timestamp.

```
-timestamp=timestamp
```

This will include the timestamp after the row key in the output:

```
-includeTimestamp=Record
```

Other Conditions

The following returns the specified number of rows per region:

```
-limit=numRowsPerRegion
```

The following will keep the specified number of rows in the cache for faster access. It also consumes more memory.

```
-caching=numRows
```

The following replaces the default delimiter with a user-specified delimiter:

```
-delim=delimiter
```

After replacing the default delimiter of whitespace, you can remove the whitespace or not. The default value is true, which removes whitespace.

```
-ignoreWhitespace=(true|false)
```

The following is used to specify the caster class to convert values (the default is Utf8StorageConverter).

```
-caster=(HBaseBinaryConverter|Utf8StorageConverter)
```

The following will fast-load data into HBase. By default it is disabled. You can set it to true. Keep in mind that sometimes it results in data loss. (See http://hbase.apache.org/ book.html#perf.hbase.client.putwal.)

```
-noWAL=(true|false)
```

The following includes the tombstone marker on the store:

```
-includeTombstone=Record
```

OrcStorage

The following allows you to replace the default stripe size of 256 MB with the new stripe size:

```
-s, --stripeSize
```

The following specifies the distance between entries in the row index:

```
-r, --rowIndexStride
```

The following sets the buffer size used for both compressing and storing the stripe:

```
-b, --bufferSize
```

The following is used to pad blocks to stripes:

```
-p, --blockPadding
```

The following sets the generic compression that is used to compress the data. Valid codec settings are NONE, ZLIB, SNAPPY, and LZO.

```
-c, --compress
```

```
employee = load 'employee.csv' using PigStorage(',') as
          (eno:int,ename:chararray,salary:int,deptno:int);
store employee into 'csvtoorc'using OrcStorage('-c SNAPPY');
```

The following allows you to specify the version of file to be written:

```
-v, --version
```

Index

© Balaswamy Vaddeman 2016
B. Vaddeman, *Beginning Apache Pig*, DOI 10.1007/978-1-4842-2337-6

▓ C

Get the eBook for only $4.99!

Why limit yourself?

Now you can take the weightless companion with you wherever you go and access your content on your PC, phone, tablet, or reader.

Since you've purchased this print book, we are happy to offer you the eBook for just $4.99.

Convenient and fully searchable, the PDF version enables you to easily find and copy code—or perform examples by quickly toggling between instructions and applications.

To learn more, go to http://www.apress.com/us/shop/companion or contact support@apress.com.

Printed in the United States
By Bookmasters